Optimizing Jet Transport Efficiency

Optimizing Jet Transport Efficiency

Performance, Operations, and Economics

Carlos E. Padilla

Illustrations by Russell Curtis

McGraw-Hill

New York San Francisco Washington, D.C. Auckland Bogotá
Caracas Lisbon London Madrid Mexico City Milan
Montreal New Delhi San Juan Singapore
Sydney Tokyo Toronto

McGraw-Hill

A Division of The McGraw·Hill Companies

pbk   1 2 3 4 5 6 7 8 9 DOC/DOC 9 0 0 9 8 7 6

Library of Congress Cataloging-in-Publication Data
Padilla, Carlos E.
      Optimizing jet transport efficiency : performance, operations, and
economics / by Charles E. Padilla.
          p.   cm.
      Includes index.
      ISBN 0-07-048208-X (pbk.)
      1. Airplanes—Jet propulsion.   2. Airplanes—Performance.
I. Title
TL709.P25   1996
629.133'349—dc20   96-16573
                                                                         CIP

McGraw-Hill books are available at special quantity discounts to use as premiums
and sales promotions, or for use in corporate training programs. For more information, please write to the Director of Special Sales, McGraw-Hill, 11 West 19th Street,
New York, NY 10011. Or contact your local bookstore.

Acquisitions editor: Shelley IC. Chevalier
Editorial team: Robert E. Ostrander, Executive Editor
                Jodi L. Tyler, Indexer
Production team: Katherine G. Brown, Director
                Rose McFarland, Desktop Operator
                Linda L. King, Proofreading
Design team: Jaclyn J. Boone, Designer   048208X
                Katherine Lukaszewicz, Associate Designer   GEN3

Contents

Acknowledgments *viii*

Introduction *ix*

1 The atmosphere and pertinent parameters *1*
  Ambient temperature *2*
  Ambient pressure *2*
  Ambient density *4*
  Effects of Mach number *4*
  Altitude *5*

2 Speeds and their measurement *7*
  Speed measurement and nomenclature *7*
  Mach to CAS conversion *10*
  Compressibility correction *11*

3 Aerodynamics *13*
  Compressible and incompressible flow *13*
  Mach number *14*
  Reynold's number *15*
  The equation of state *16*
  Aerodynamic parameters *17*
  Pressure distribution over an airfoil *18*

4 Airfoils and wings *21*
  Airfoil properties *21*
  Wing properties *22*
  Lift *23*
  Drag *24*

5 Propulsion *27*

Turbojets and turbofans *27*
Engine stations *27*
Thrust-setting parameters (TSP) *29*
Fuel flow (*Wf*) *31*
Thrust ratings *31*
Thrust limitations *34*

6 General performance topics *37*

Thrust available *37*
Thrust required *37*
Bank-angle influence on G-loads (load factor) *40*
Buffet limits *41*
Bank-angle influence on turn radius *47*
Bank-angle influence on climb gradient *49*
Radius of turn in the presence of winds *54*
V-n diagrams *58*
Maximum speeds *61*

7 Takeoff *63*

Takeoff speeds *64*
Runway configuration *66*
The earthbound portion of the takeoff *68*
The airborne portion of the takeoff *81*
Other takeoff limitations *93*
Takeoff equations *95*

8 Climb *99*

Climb equations *99*
Rate of climb *100*
Angle of climb and climb gradient *109*
Climb performance with one engine inoperative *115*
Economy climb *120*

9 Cruise *127*

Force-speed diagrams *127*
Specific range *131*
Step climb *142*
Short trip optimum altitude (STOA) *143*

Integrated range   *144*
Integrated time   *147*
Point of no return (PNR)   *147*
Tankering   *149*
Range and endurance formulations   *150*

10 Economy cruise   *153*
Cost index (CI)   *154*
Economy speeds   *155*
Optimal trajectories   *162*

11 Descent and landing   *163*
The effects of thrust   *164*
The effect of drag   *164*
The effect of speed   *165*
Drift-down   *165*
Holding   *167*
Equations of motion   *168*
The effect of weight on *ROD*   *170*
Landing   *172*

12 Weight and balance   *177*
Operational allowances and the curtailed envelope   *178*

13 Payload-range curves   *187*
Payload-fuel curves depict limited resources   *187*
Development of payload-range curves   *188*
Equipment evaluation   *193*

Abbreviations   *195*

References   *199*

Index   *201*

About the author   *205*

Acknowledgments

Among many individuals, the flight operations engineering departments at Boeing and at America West Airlines are salient for their contributions to my education throughout the years in the subject of aircraft performance. Their professionalism and vast technical expertise are highly appreciated.

But after putting all the materials together, only the patience and understanding exhibited by the editors and staff at McGraw-Hill made it possible to transform my manuscript into a publishable book. My thanks to Acquisitions Editor Shelley IC. Chevalier and everyone else who worked on the book.

My thanks also to Russell Curtis for making the figures.

Introduction

This book has its origin in a set of notes prepared for a seminar on aircraft performance. It is a compilation of some of the concepts and tools used by performance engineers in the analysis of airline-related performance problems. Although the book does not present the totality of all the tools available, it does deal with enough of them to allow the beginning student and the practicing engineer to derive much benefit from its pages.

Not only the student new to the problems of aircraft performance, but people involved in the day-to-day performance problems of a fleet of jet transports will benefit from this book. Likewise, the book will be of value to anyone interested in aircraft performance within the stated context: pilots, avionics software engineers, airline dispatchers, plus aircraft and avionics marketing personnel, as well as aircraft design engineers.

Often in the airline environment, one finds a lack of understanding of the dynamics and economics of flight and the capabilities and limitations of the modern jet transport. This book addresses these areas drawing from all pertinent disciplines: aerodynamics, propulsion, mechanics, as well as the governing regulations in the United States (Federal Aviation Regulations).

The layout of the book is such that individuals of different background and mathematical sophistication can benefit from its contents. With an interest in presenting material useful to all, wherever possible, the more mathematical presentations have been relegated to later parts of the chapter for the benefit of those who are interested in the more technical details without sacrificing usefulness for the other readers. Nevertheless, when selected treatments are necessarily mathematical in character, the developments are presented as clearly as possible.

Of particular interest to pilots will be the sections dealing with takeoff performance and cruise performance. The takeoff problem, as presented in this book, is far deeper than needs to be mastered by most crewmembers; indeed, the presentation attempts to explain how the charts in the aircraft flight manual are conceived. Rather than mastery, a pilot will benefit from understanding the different parameters involved in the calculation of a takeoff, as well as the applicable limitations.

Likewise, pilots will gain additional understanding of the cruise problem from the presentation in the book. The contents of chapter 9 will help to dispel such misconceptions as long-range cruise being the most fuel efficient speed schedule; it will also contribute to understand better the concepts of optimum altitude and step climb.

There are many good sources of material on aerodynamics and propulsion; consequently, the presentation of these subjects here is minimal, intended only to introduce some basic concepts and terminology. The reader that is interested in more in-depth treatment is encouraged to consult any of the many good texts available on these subjects.

The reader who intends to use this work as a textbook should be comfortable with the basics of statics and dynamics, as well as have a mastery of the fundamental concepts of algebra, trigonometry, and some elementary calculus. As a textbook, the work would fit well at the second-year level of a standard engineering curriculum. Again, expertise in aerodynamics and propulsion is not required, but it would certainly enhance the process of learning some of the concepts presented here.

A considerable effort has been dedicated by the author to accuracy and clarity of presentation. Not claiming infallibility, the author welcomes any comments and suggestions that may be incorporated in future editions of this work.

1

The atmosphere and pertinent parameters

Although the actual atmosphere reaches heights of a few hundred thousand feet, it has been divided somewhat arbitrarily into a few major portions, within which the pressure and temperature variations have been defined by the world's aeronautical community (the International Civil Aviation Organization, ICAO); the result is known as the International Standard Atmosphere (ISA).

For the purpose of studying the performance and dynamics of existing transport aircraft, the layer of air to be considered is confined to altitudes between 1,000 ft. below sea level (SL) and 60,000 ft. above SL. The lower altitude becomes important when considering flight at very low altitudes or in high atmospheric pressure conditions, when the pressure altitude might be less than sea level.

The *international standard atmosphere* (ISA) assumes the following base values at a latitude of N45°32'40":

pressure: p_0 = 2,116.22 lb/ft²;

density: ρ_0 = 0.002377 slugs/ft³;

temperature: T_0 = 59°F = 518.688°R = 15°C = 288.16°K;

gravity: g_0 = 32.1741 ft/sec²;

Gas constant: R = 53.35 ft/°R.

In the study of aircraft performance and atmospheric flight dynamics, the atmosphere is characterized by the use of several parameters as will be explained below. The use of these parameters will become evident in the treatment of subjects presented later in the book.

Ambient temperature

For convenience, rather than as an accurate description of the actual condition, the atmosphere has been divided into layers through which either the temperature or the temperature variation with altitude is constant. Such a temperature model can be described as follows:

Altitude	Temperature (°R)
SL to 36,089	$-3.566*(h/1{,}000)$
above 36,089	$389.988°$

We are now in position to define the temperature ratio, T_{amb}/T_{SL}:

$$\frac{T_{amb}}{T_{SL}} = \theta = 1 - (6.8753 \times 10^{-6})h_p \qquad (1\text{-}1)$$

for $h_p < 36{,}089$ ft; and

$$\theta = 0.7519 \qquad (1\text{-}2)$$

for $h_p \geq 36{,}089$ ft;

where h_p refers to the pressure altitude, a concept that will be explained later in this chapter.

Clearly it is seldom that the conditions described by the ISA are found in practice. To address these off conditions, the concept of delta ISA (ΔISA) was introduced. ΔISA refers to the temperature shift between the theoretical ISA and the actual conditions; this shift is usually given in degrees Celsius, °C. Under this convention, for example, a +40-degree day will exhibit a shift of +40°C throughout the entire temperature distribution of the atmospheric layers, including the temperature at sea level (Fig. 1-1). θ then becomes,

$$\theta = \frac{T_{amb} + \Delta ISA}{T_{SL} + \Delta ISA} \qquad (1\text{-}3)$$

It is important to recognize that temperature can refer to static or ambient conditions. Static temperature, T_s (also known as *static air temperature*, SAT), exists immediately next to the skin of the moving body, whereas ambient temperature, t_{amb} (aka *outside air temperature*, OAT), is the prevailing temperature at some distance from the body. Unfortunately, it is not uncommon for both of these to be used interchangeably, even though T_s responds to increasing flow velocity by decreasing, exhibiting a behavior similar to that of p_s.

Ambient pressure

Through the use of fluid mechanics and the temperature model of the previous section, it is possible to arrive at an expression describing the ambient pressure as a function of pressure altitude:

$$\delta = (1 - 6.88 \times 10^{-6} h_p)^{5.26} \qquad (1\text{-}4)$$

for $h_p < 36{,}089$ ft; and

$$\delta = .223360 e^{\left(\frac{36{,}089\, -\, hp}{20{,}805.7}\right)} \qquad (1\text{-}5)$$

for $h_p \geq 36{,}089$ ft;

where δ is the ratio of local ambient pressure to standard sea-level ambient pressure:

$$\delta = \frac{p_{amb}}{p_{SL}} \qquad (1\text{-}6)$$

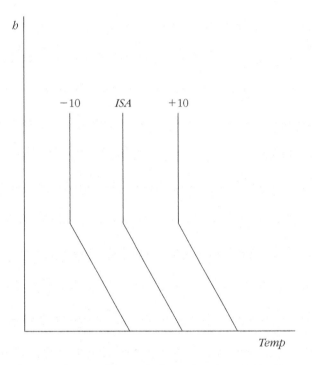

Fig. 1-1 ΔISA *refers to shifts in the ISA temperature profile.*

As is the case with temperature, there are two pressures: static pressure, p_s, and ambient pressure, p_{amb}. As the air moves around a body, the pressure around the body drops in response to the increased local speed of the flow; the resulting pressure at each point is called the static pressure. In contrast, the pressure existing in the neighborhood of the moving body, but far enough away not to be affected by its motion, is the ambient pressure; therefore, the ambient pressure is generally higher than the static pressure.

Aircraft speed and altitude measurement are accomplished in part by measuring static pressure through the static pressure port. Since this reading is intended to be ambient pressure, some correction is necessary. This correction is one of the important parts of the flight test program of an aircraft and is published in the certification document, the airplane flight manual (AFM). Modern jet transports electronically incorporate this correction in their instruments.

Ambient density

Another ratio used in the study of aircraft performance is the density ratio, σ, which again relates local ambient conditions to sea-level ISA conditions,

$$\sigma = \frac{\rho_{amb}}{\rho_{SL}} \qquad (1\text{-}7)$$

where ρ is the air density. Also, from thermodynamics, it can be shown that

$$\sigma = \frac{\delta}{\theta} \qquad (1\text{-}8)$$

Effects of Mach number

Pressure, density, and temperature are affected appreciably by the compressible character of air at relatively high speeds (higher than $\approx$200 fps). When considering flight at these speeds, the concepts of total pressure and total temperature become useful,

$$p_t = p_s [1 + 0.2M^2]^{3.5} \qquad (1\text{-}9)$$

$$T_t = T_s [1 + 0.2M^2] \qquad (1\text{-}10)$$

where M is the Mach number. Total temperature, T_t, is also known as *total air temperature*, TAT, or *stagnation temperature*. Likewise, the total pressure is sometimes called *stagnation pressure*.

These parameters, in turn, can also be referred to *standard conditions*, where the resulting expressions are:

$$\delta_t = \frac{p_t}{p_{SL}} = \frac{p_s}{p_{SL}} [1 + 0.2M^2]^{3.5} \qquad (1\text{-}11)$$

$$\theta_t = \frac{T_t}{T_{SL}} = \frac{T_s}{T_{SL}} [1 + 0.2M^2] \qquad (1\text{-}12)$$

TAT is sometimes mistaken for *ram air temperature* (RAT). RAT is the temperature at the recovery end of the temperature probe. It differs from TAT by the magnitude of the installation error. RAT can be thought of as local TAT at the tip of the probe.

Altitude

There are at least four different interpretations of altitude:

- Pressure altitude, which, as the name implies, depends on the pressure measurement at the point of interest
- Geometric altitude, which corresponds to the physical distance existing between the aircraft and the chosen reference
- Density altitude
- Geopotential altitude

Pressure altitude really measures a difference in pressure between the reference surface (sea level, for example) and some other surface, or layer of air, throughout which the pressure is constant; whereas geometric altitude is the distance separating two parallel surfaces, one being the reference surface and the other one located some distance above it.

Conventionally, the reference layer is chosen as the *mean sea level* (MSL). If, for some reason, there is interest in referring to another datum on the surface of the Earth, the reference is made to an altitude *above ground level* (AGL). The latter is frequently the case in the context of obstacle clearance problems.

Although identical at first sight, pressure and geometric altitudes differ. As an illustration, consider a layer of air at 20,000 ft. above the ocean extending over an area of 10,000 square miles. Under standard atmospheric conditions, the geometric and pressure altitudes are both 20,000 ft. Now suppose that there is a tropical depression over a section of 1,000 square miles of the original area. Over this portion of the study area, the pressure has dropped below the standard, and, to someone measuring altitude within the layer with a barometric instrument, the reading would indicate an altitude of perhaps 21,000 ft. (tropical depression => lower pressure => higher altitude), when in reality the distance separating the instrument from the ocean is still 20,000 ft.

A third interpretation of altitude is density altitude. The term density altitude is often encountered in the warm summer months, when the heat of the day makes the density of the air decrease to a value that, under standard conditions, would be associated with higher altitudes. Density altitude is rarely used in performance work done by fleet operators, where most of the parameters influenced by atmospheric conditions can be handled using pressure and temperature ratios, δ and θ.

Occasionally, density altitude is important in performance work, and is determined from the prevailing ambient pressure and temperature through the *equation of state* (Equation 3-5). Accordingly, the ISA has a standard density schedule that can be calculated from the pressure and temperature values at each altitude. For nonstandard conditions, a nonstandard atmosphere, the nonstandard density schedule will be offset from the standard density schedule. As an illustration, consider that at 10,000 ft. the standard density is .001755 slug/ft^3. Now, in a hot day, the density might be .001596 slug/ft^3, which corresponds to an altitude of about 13,000 ft. in the standard atmosphere; we say then that, under this particular hot-day condition, the density altitude is 13,000 ft., not the normal 10,000 ft.

Another altitude often found in the literature is *geopotential altitude*. It is defined in terms of the distance existing between the center of the Earth and parallel surfaces around the globe. The definition is such that, over the entire extension of one of these surfaces, the gravitational potential remains the same. In the study of aircraft performance problems, the variations in altitude do not result in a large variation in gravitational potential; therefore, the concept of geopotential altitude is not used often.

Variations in gravity potential are nevertheless often taken into account in aircraft condition monitoring, when an aircraft will be subject to a detailed assessment of its performance characteristics while in revenue flight.

In summary, the following altitudes may be considered:

- Pressure altitude: An altitude defined in terms of pressure differential with respect to sea level.
- Geometric altitude: An altitude defined in terms of the vertical separation from sea level.
- Density altitude: An altitude defined in terms of the difference in density with the ISA.
- Geopotential altitude: An altitude defined in terms of gravity potentials.

2

Speeds and their measurement

The realm of aircraft performance involves the assessment of the dynamic state of the aircraft. Speed, a measure of the kinetic energy of the aircraft, is therefore of primary importance.

The measurement of speed is usually a straightforward process when dealing with objects that remain fixed to the ground. The reason is simple: The ground neither moves nor deforms. In measuring aircraft speed, we are faced with the problem that the medium within which aircraft move both deforms (compresses) and is in motion (winds). To solve the problem of describing and measuring the speed of aircraft, a number of artifacts have been devised.

Speed measurement and nomenclature

Probably the best known instrument used in speed measurement is the *pitot-static* system, shown schematically in Fig. 2-1.

This device simply measures the difference in pressure that exists between its tip and side ports. At the tip, the existing pressure is total, or *stagnation pressure*, p_t. The side port is exposed to *static pressure*, p_s. The difference between these two in incompressible flow is the dynamic pressure, q, which is directly proportional to the square of the local velocity,

$$q = p_t - p_s = \tfrac{1}{2}\rho V^2 \qquad (2\text{-}1)$$

Under standard sea-level conditions, at speeds low enough not to cause the air mass to compress appreciably, the speed measured by such an apparatus is considered to be the *true airspeed* (TAS).

The flight regimes of interest here involve high speeds and high altitudes. Both of these parameters affect the density of the air and therefore have a distorting effect on the attempted speed measurement. The speed measuring equipment in modern aircraft is calibrated such that TAS is correctly sensed only at standard

7

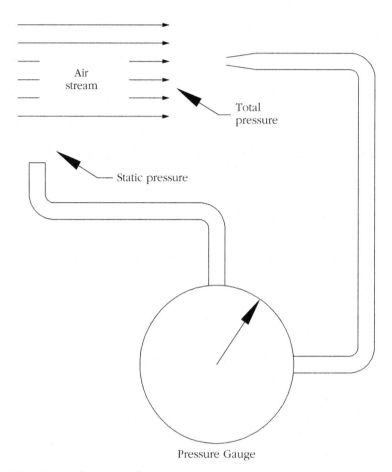

Fig. 2-1 *Schematic of a pitot-static system.*

sea-level conditions. The distortions in the measurement introduced by altitude and the combined effect of compressibility with altitude must be considered.

First, as the altitude increases, the impact pressure sensed at the front of the pitot tube will be due to less-dense air, the speed thus measured will be under-estimated. This is easily solved by considering the effect of density variation with altitude.

Second, at higher altitudes, if the speed is high enough to cause the air to compress appreciably, the impact pressure felt by the pitot tube will be due to an air density that is higher than the actual ambient density by virtue of the compression that the air has undergone. If the altitude at which the measurement is done is sea level, there is no error incurred because the instrument was calibrated for sea level, incompressible flow. At other altitudes, an error is introduced that is due to the combined and convoluted effect of both altitude and compressibility that cannot be removed by only accounting for the different density; the effect of high speed is also embedded in the error and must also be included in the correction. A development of this correction term can be found in any basic text on aerodynamics.

Figure 2-2 might help in understanding the relationship between the different speeds involved in speed measurement. Note in Fig. 2-2 that at sea level CAS = EAS = TAS since the compressibility-at-altitude error is nonexistent, as is the error due to altitude associated with EAS.

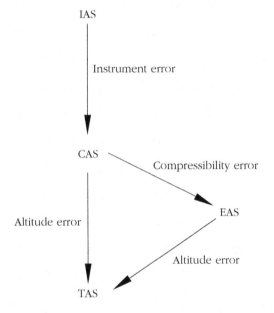

Fig. 2-2
Relationship between speeds.

The following nomenclature is thus used in performance work:

- V_i or IAS—Indicated airspeed is the speed presented for reading by the crew at the cockpit. This speed is usually not corrected for error arising from the particular location of the pitot-static system and therefore must still be subject to what is called *position error calibration.*

- V_c or CAS—Calibrated airspeed is the result of correcting IAS for position error, $V_c = V_i + \Delta V_p$, where ΔV_p is the position error correction. In the more sophisticated systems, the speed reading presented to the crew has been corrected for position error, in which case IAS = CAS. More commonly, the position error correction is presented in the AFM.

- V_e or EAS—Equivalent airspeed is the same as CAS only at SL; otherwise, EAS and CAS differ by the compressibility error, which depends on the combined effect of both altitude and Mach number, $V_e = V_c - \Delta V_c$. EAS is less than CAS because, as stated earlier, the instrumentation is fooled into perceiving a higher than real velocity due to the increased density of air at its total pressure port. The compressibility correction, ΔV_c, will be developed in a later paragraph.

- V_T or TAS—True airspeed is the actual relative speed between the aircraft and the air mass. The difference between TAS and EAS is due to the difference in density between SL and the altitude where the speed measurement is made. EAS must be corrected for an altitude error to convert it to TAS:

$$TAS = EAS \sqrt{\frac{\rho_{SL}}{\rho}} \tag{2-2}$$

Often the above speeds are given in knots and abbreviated accordingly: KTAS, KIAS, KEAS, KCAS.

From Equation 1-9, it is evident that there is a 1-to-1 relationship between M^2 and the ratio of p_t/p. A modern airplane's electronic sensors and instrumentation can calculate M directly based on this unique relationship. With M and the ambient temperature, it is straightforward to calculate TAS. Nevertheless, the more complex calculation that yields IAS is still needed, as IAS is one of the parameters used in establishing compliance within the ATC environment.

Mach to CAS conversion

In the following development it will be assumed that the instrument and position errors have been eliminated and CAS = IAS. Revisiting Equation 1-9 and rewriting it in a slightly modified form yields

$$p_t - p = p\,[(1 + .2M^2)^{3.5} - 1] \tag{2-3}$$

Under standard SL conditions, TAS = CAS. For these conditions we can use CAS in the substitution for M in the above equation:

$$p_t - p_0 = p_0 \left[\left(1 + .2\left(\frac{CAS}{a_0}\right)^2\right)^{3.5} - 1\right] \tag{2-4}$$

At this point, we consider the following questions, remembering that δ, the atmospheric pressure ratio, characterizes pressure altitude:

- At sea level, under standard conditions, what value of CAS will generate the same pressure difference ($p_t - p$) as that generated by flying at a given M, at a given higher altitude?
- For an aircraft flying at altitude, at some M, how fast would the aircraft have to fly at SL to generate the same pressure rise?

These two questions are really the same, except posed from different perspectives. One way to reconcile these perspectives is to set Equations 2-3 and 2-4 equal to each other. After some algebraic manipulation, the result is:

$$M^2 = 5\left[\left(\frac{1}{\delta_{amb}}\left(\left[1 + .2\left(\frac{CAS}{a_0}\right)^2\right]^{3.5} - 1\right) + 1\right)^{\frac{1}{3.5}} - 1\right] \tag{2-5A}$$

This expression of M in terms of CAS and δ, and its counterpart of CAS in terms of M and δ, are very often used in converting M to CAS and vice versa. Solving for CAS,

$$CAS = a_0\sqrt{5}\,\sqrt{[\delta[(1 + 0.2M^2)^{3.5} - 1] + 1]^{\frac{1}{3.5}} - 1}\qquad(2\text{-}5B)$$

These expressions are valid for subsonic flight only. In supersonic flight, the pressures sensed by the measuring instruments reflect the conditions behind a shock wave and do not correspond to the true speed of the aircraft.

Compressibility correction

The compressibility correction is the difference between EAS and CAS at the same condition of flight. One way to express the condition of flight is to express it in terms of the pressure rise experienced by the Pitot-Static system,

$$p_t - p$$

If expressions can be found that relate both CAS and EAS to this pressure differential, it will then be possible to find the compressibility correction.

Substituting the following in Equation 2-3

$$M = \frac{V}{a}\qquad(2\text{-}6)$$

$$a = \sqrt{\gamma\frac{p}{\rho}}$$

and then multiplying and dividing by ρ_0 yields:

$$p_t - p = p\left[\left(1 + 0.1429\frac{\rho_0}{p}\,V^2\sigma\right)^{3.5} - 1\right]\qquad(2\text{-}7)$$

Setting Equations 2-4 and 2-7 equal to each other, and solving for V_e yields

$$V_e = V\sqrt{\sigma} = a_0\sqrt{5\delta\left[\left(\frac{\left[1 + .2\left(\frac{CAS}{a_0}\right)^2\right]^{3.5} - 1}{\delta} + 1\right)^{.2857} - 1\right]}\qquad(2\text{-}8)$$

V_e (= EAS) can be evaluated for various combinations of CAS and altitude, from which the difference EAS − CAS can be calculated and tabulated, such as shown in Fig. 2-3.

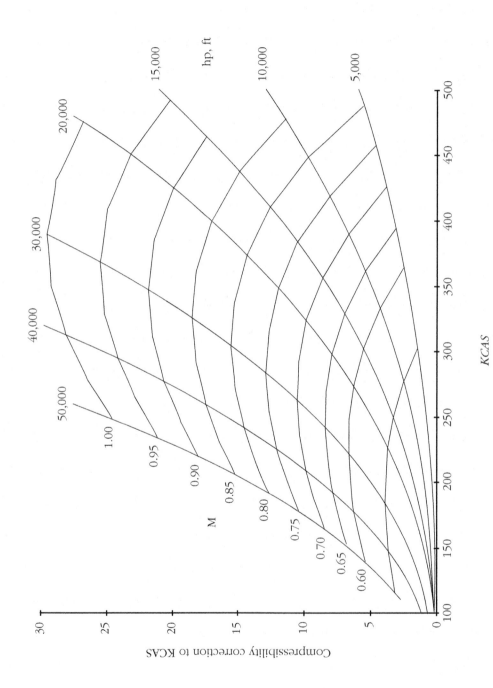

Fig. 2-3 *Compressibility correction.*

3

Aerodynamics

Aerodynamics is a specialization of the wider field of fluid mechanics. Generally speaking, in aerodynamics we are concerned with the study of fluids that are susceptible to compressibility, and in this book we will deal only with air.

Numerous valuable books deal with the basic concepts of aerodynamics (see References); therefore, in this presentation only some of the salient points of this field that are necessary to build an understanding of the aircraft performance problem will be treated in some detail. The reader may choose to delve more deeply into the subject by consulting some of the references cited.

Compressible and incompressible flow

The realm of incompressible flow is characterized by airspeeds much lower than the average speed of the air molecules; at standard temperature and pressure (STP) conditions, speeds of roughly 200 fps or less are considered to be in the regime of incompressible flow. Most general aviation operations (GA) and most ground vehicle operations fall under this category.

To study the correlation between flow parameters, aerodynamicists resort to the Bernoulli equation, a representation of the energy balance existing in the flow. Due to the different nature of compressible and incompressible flows, the Bernoulli equation has a different form for both of these flow regimes.

For incompressible flow:

$$p_1 + \frac{1}{2}\rho V_1^2 = p_2 + \frac{1}{2}\rho V_2^2 = p_{t\,inc} \tag{3-1}$$

For compressible flow:

$$\frac{\gamma_1}{\gamma-1}\frac{p_1}{\rho_1} + \frac{V_1^2}{2} = \frac{\gamma_2}{\gamma-1}\frac{p_2}{\rho_2} + \frac{V_2^2}{2} = p_{t\,comp} \tag{3-2}$$

where,

p_n = static pressure at station n, psf
p_t = total pressure, psf
V_n = air velocity at station n, fps
γ_n = ratio of specific heats at station n
ρ_n = air density at station n, slugs per ft^3

13

The equations relate the flow at different locations, as shown in Fig. 3-1. Note that the constant p_t indicates that the total pressure remains constant; another way of expressing the same idea is to say that the total energy in the flow is constant.

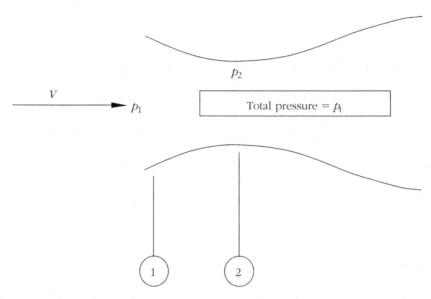

Fig. 3-1 *Reference flow and station designations for p_1 and p_2.*

In the case of incompressible flow, the total energy is the sum of the potential energy (represented by the static pressure) and the kinetic energy (represented by the dynamic pressure). As the flow velocity varies, there will be a balance of energy between kinetic and potential such that the total pressure remains constant.

Similarly, in the compressible equation, there is a balance of energy between kinetic and potential with the addition of elastic energy represented by the different values of ρ, which is the air density. Strictly speaking, *elastic energy* is potential energy stored within the structure of the air mass. We highlight it here to underscore the notion that there is an additional contribution to energy storage that is not considered in incompressible flow.

The quantity γ is the ratio of the specific heats of air. Air at temperatures of normal flight has a γ of 1.4. For more details on the nature of specific heat, refer to any standard thermodynamics text.

Mach number

In aeronautical work, we are faced with problems of motion in regimes of flight that vary widely in speed, size of the object, ambient conditions, and forces acting on the object. To deal effectively with the behavior of all these parameters, aerodynamicists have developed tools that simplify their work by establishing ratios of the different forces acting on the object under study.

As an object moves through a mass of air, it compresses this mass (negligibly below 200 fps), and pushes the molecules ahead of it, much as a boat creates a bow wave as it moves through the water surface. These molecules have an inherent ability to move forward, away from the intruding object, which is a direct function of the kinetic energy possessed by the molecules. A measure of this molecular kinetic energy is known commonly as *temperature*. The higher the air temperature, the more ability the molecules have to move under the influence of a penetrating object, and the more elasticity the air exhibits to objects moving within it.

We see here the presence of two concepts that play an important role in the study of aerodynamic phenomena: the elasticity of the air and the inertia of the objects moving within it.

Since sound waves are nothing more than periodic manifestations of the elasticity of the medium within which they propagate, a good measure of air's elasticity is the speed of sound within it. Moreover, a good representation of the inertia of a body is its speed. If we are interested in characterizing the relationship between the elasticity of the air and the motion of a body that taxes this elasticity, we form the ratio

$$\frac{V}{a} = Mach\ Number = M \qquad\qquad (3\text{-}3A)$$

where,

V = speed of the object, fps

a = speed of sound, fps

The velocity of sound at any altitude can be expressed in terms of its sea-level value through the use of the temperature ratio, θ,

$$velocity\ of\ sound\ @\ h = a_h = a_{SL}\sqrt{\theta_h}$$

Mach number can then be expressed as

$$M = \frac{V}{a_o\sqrt{\theta_h}} \qquad\qquad (3\text{-}3B)$$

where V is the true airspeed of the object.

Reynold's number

A particularly important set of conditions exists in the layers of air that are in contact with the moving object itself (Fig. 3-2). Here there are molecules whose speed must match exactly that of the moving object, whereas the molecules that are farther from the object's surface are affected less by the object. The region of the air mass that is sheared by the presence of the moving object is called the *boundary layer*. Note that this is not the same as the region of the air mass that

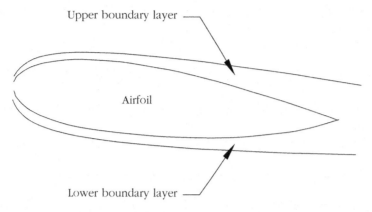

Fig. 3-2 *Schematic of the boundary layer on an airfoil.*

is affected by the pressure wave generated by the moving object; this latter region extends much farther.

In close proximity to the moving object, therefore, we have a region of air whose viscosity determines how much shearing force it takes to move the adjacent layers of air. We are now concerned with shearing forces and inertia forces. The ratio of interest is then

$$\frac{INERTIA\ FORCES}{SHEARING\ FORCES}$$

or,

$$\frac{\rho Vl}{\mu} = Reynold's\ Number = RN \tag{3-4}$$

where

V = air velocity, fps
l = reference dimension, ft
μ = viscosity, lb-sec per ft^2
ρ = air density, slugs per ft^3

The equation of state

The relationship between the pressure, temperature, and density of a gas is important in the study of aerodynamic phenomena. This relationship is developed by studying the thermodynamic characteristics of air and establishing what has come to be known as the *equation of state*,

$$p = \rho gRT \tag{3-5}$$

where

p = pressure, psf
ρ = density, slugs per ft^3

R = gas constant = 53.4
T = absolute temperature, °R
g = gravitational constant = 32.2

The equation of state is a direct result of Charles's and Boyle's laws (Ref. 2). It describes the state of a perfect gas in terms of the prevailing ambient conditions. Although air is not a perfect gas, its behavior can be approximated sufficiently accurately by making this assumption.

The importance of the equation of state is that it allows us to study any one of three quantities (pressure, temperature, and density) in terms of the other two.

Aerodynamic parameters

In the study of aerodynamic phenomena, the use of nondimensional parameters is common. Nondimensional parameters permit the study to be generalized; for example, to study the forces acting on a wing in flight, a scale model of the wing can be constructed and subjected to reduced speeds and loads, thus reducing the complexity of the study.

The following parameters will be used in developing the material in this presentation:

Lift coefficient:

$$C_L = \frac{L}{\frac{1}{2}\rho V^2 S} \tag{3-6A}$$

or, alternatively,

$$C_L = \frac{L}{(1481.351)\,\delta\,M^2 S} \tag{3-6B}$$

where,

L = lift, lbs
ρ = air density, slugs per ft^3
V = airspeed, fps
S = reference area, ft^2
δ = ambient pressure ratio
M = Mach number

Drag coefficient:

$$C_D = \frac{D}{\frac{1}{2}\rho V^2 S} \tag{3-7A}$$

or, alternatively,

$$C_D = \frac{D}{(1481.351)\,\delta\,M^2 S} \tag{3-7B}$$

where,

  D = drag, lbs

Moment coefficient,

$$C_m = \frac{M}{\frac{1}{2}\,\rho\,V^2\,Sl} \tag{3-8A}$$

or, alternatively,

$$C_m = \frac{m}{(1481.351)\,\delta\,M^2\,Sl} \tag{3-8B}$$

where,

  m = moment, ft-lbs
  l = reference length, ft

   The forms using M and δ can be derived from those using V and ρ through the use of thermodynamic transformations known as the *isentropic relations.*

   Other coefficients are necessary in more-detailed presentations, such as stability/control and aeroelasticity, but for the purposes of the material treated in this book, the coefficients presented above will suffice.

Pressure distribution over an airfoil

Figure 3-3 depicts an airfoil immersed in an airstream at an angle α. Also shown is a representation of the pressure distribution over the airfoil. Due to the curvature over the upper surface, the air is forced to increase its speed from the free-stream velocity V_∞, to a higher value V_u, which in turn provokes a decrease in pressure from p_{amb} to p_u. Similarly, the lower surface presents a similar set of circumstances, but, since the curvature is less pronounced, the associated rise in velocity and drop in pressure are also less pronounced.

   Because the pressure on the upper surface is less than the pressure in the lower surface, the net effect is the lift vector shown. Note that the lift vector is perpendicular to the free stream velocity vector and drag parallel to it; as a result, an increase in α has an associated increase in lift and drag, and possibly a tilting of the total aerodynamic force vector.

   Figure 3-4 shows a plot of the effect of angle of attack, α, on C_L in the normal flight regime and shows also the point at which stall can be expected. At that point, the flow separates from the upper surface, leaving a region of essentially ambient air pressure, which is insufficient to produce the necessary lift (Fig. 3-5). The flow separation that ensues at high α values is responsible for the vibration, or shaking, associated with stall.

   Figure 3-6 shows another source of separation, namely a shock wave on the surface of the wing, produced by a local Mach number higher than one. Such a phenomenon can occur at high subsonic Mach flight during which the local flow over the wing accelerates past M = 1. In this regime of flight, drag begins to increase rapidly, giving rise to the drag divergence region.

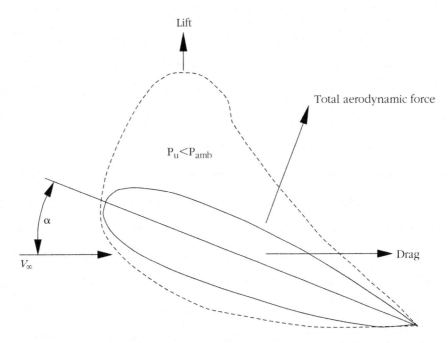

Fig. 3-3 *Pressure distribution (dashed line) over an airfoil.*

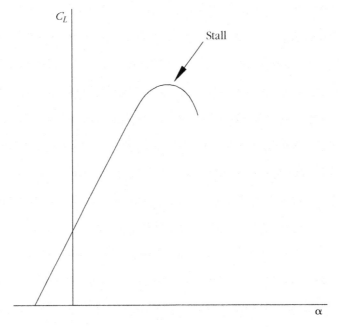

Fig. 3-4 *Variation of lift coefficient with angle of attack.*

Fig. 3-5 *Schematic of flow separation over an airfoil.*

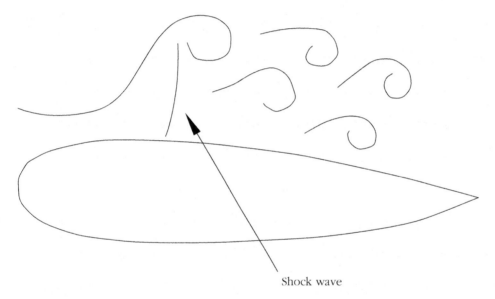

Fig. 3-6 *Schematic of flow separation due to shock wave.*

4

Airfoils and wings

The modern jet transport has been designed to perform efficiently during cruise; as a consequence, a transport does not perform as efficiently in other flight regimes. The aircraft designer is therefore responsible for the difficult task of maximizing cruise performance and, at the same time, minimizing any penalties that might arise from flight at off-design conditions. The most important tools that designers have at their disposal for this purpose are wings and the airfoils of which these wings are comprised.

The sophistication of wing and airfoil design must then be matched by an equally sophisticated operation. A thorough treatment of airfoils and wings can be found in other specialized works; we will limit ourselves here only to pointing out some important concepts that bear mentioning.

Airfoil properties

When reference is made to an airfoil section, or simply "airfoil," we refer to the shape of the cross section of a wing or any other device used to generate lift. Figure 4-1 is a cambered airfoil along with some of the concepts used when dealing with airfoils:

- The *leading edge* (LE) is the edge of the section that faces the oncoming flow.
- The *trailing edge* (TE) is the edge of the section away from the oncoming flow.
- The *chord* of the section is the straight line that joins the leading edge with the trailing edge.
- The *meanline* is the locus of points that lie midway between the upper and lower surfaces of the airfoil.
- *Camber* is the maximum distance between the meanline and the chord. In a symmetrical airfoil, the camber and meanline coincide; therefore, the airfoil has no camber, and both the upper and lower surfaces are identical.
- *Angle of attack* is the angle subtended between the oncoming flow and the chord.

- *Leading-edge devices* are also sometimes called *slats*. They are aerodynamic surfaces—not necessarily airfoil-shaped—that are installed at the leading edge of the airfoil and augment its capability to generate lift.
- *Trailing-edge devices* are more commonly known as *flaps*. They are installed at the trailing edge of the airfoil, and, like slats, augment the airfoil's capability to produce lift.
- *Spoilers* are devices installed on either the upper or lower surface, or both, to produce drag and inhibit lift.

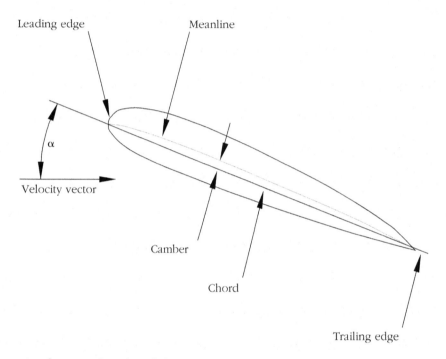

Fig. 4-1 *Cambered airfoil nomenclature.*

Wing properties

A wing can be made up of many airfoil sections, varying continuously from the root to the tip. This variation is dictated by the need to optimize the ability of the wing to generate lift from locally varying flows since the flow impacting the wing at its root is different from the flow found at the tip.

A number of terms are important to understand wing design terminology:

- *Span* is the straight-line distance from one wing tip to the other.
- Wing *area* is the area of the shadow that the wing casts when illuminated from above. Indeed, the area of the wing is quite an arbitrary concept. Because the area of the wing is used in performance

work for no other purpose than as a reference, any value that
approximates the actual geometrical area will suffice, as long as the
value used is always the same.

- The *mean aerodynamic chord* (MAC) is the result of dividing the wing
  area by the span. The MAC is another parameter used in reference work.
- *Aspect ratio* (AR) is the result of dividing the square of the span by the
  area. The AR is a measure of the wing's fineness. Its use is of more
  importance in pure aerodynamic work, not so much in performance.
- The *quarter chord* is the locus of points that are 25 percent of the
  distance from the LE to the TE.
- *Taper* is the ratio between the root chord and the tip chord.
- *Sweep* is the angle between a line perpendicular to the fuselage and the
  quarter chord line.

Lift

The primary aerodynamic function of all airframes is either the generation of lift
or carrying payload while minimizing drag. The degree to which the airframe
can generate lift while minimizing the inevitable drag and aerodynamic moments is determined by the design and intended use of the aircraft.

Lift is generated by the pressure differential that exists between the upper
and lower surfaces of the wing. This differential is the result of the higher velocity flow that exists on the upper surface of the wing and the relatively lower
velocity flow along the lower surface of the wing. From Bernoulli's principle, we
know that lower static pressures are associated with higher velocity flows.

Three sources of lift in the wing are:

- *Chord length.* The longer the chord, the more surface area over which a
  pressure differential can act, but also the more surface over which
  friction drag is generated.
- *Angle of attack* (α, alpha). A more pronounced alpha will induce higher
  pressure differentials between the upper and lower surfaces of the
  wing. The limit, of course, is the alpha at which the flow separates; at
  that value, lift is no longer produced, and a stall condition ensues.
- *Camber.* Like angle of attack, camber has the effect of inducing higher
  pressure differentials for a given surface area. Excessive camber
  produces unnecessary drag and possibly unwieldy dynamic behavior.

Indeed, when an aircraft is in a regime of flight that requires the maximum
generation of lift, all of the above items are employed by deploying flaps (increased chord and camber) and increasing the angle of attack. Note that the lift
generated by an aircraft in approach configuration—when most of the available
flaps/slats will be deployed and the angle of attack is relatively high—is the
same as the lift generated by the same aircraft in a cruise condition, at minimum
alpha, and in a clean configuration. The weight of the aircraft might be the same
under both configurations, and this weight has to be balanced with the lift gen-

erated by the wing, but at the lower speed, the lift has to be generated from a lower energy air mass, leaving it up to the aerodynamic devices to extract the most from the available air flow.

In chapter 3, the concept of lift coefficient was introduced. Recall that C_L is a nondimensional parameter used in the study of the aerodynamic characteristics of a wing. It can also be given a more immediate interpretation as a measure of the lift potential of the wing in a particular configuration. At a low alpha and without flaps, for example, the wing will have a lower C_L (lower lift potential) than at high alpha and with flaps. This potential is translated into lift when C_L is multiplied by the quantity $\frac{1}{2}\rho V^2 S$ (see Equation 3-6A).

In pursuit of flight at lower speeds, consider what happens to Equation 3-6A when solved for velocity:

$$V = \sqrt{\frac{2L}{\rho C_L S}}$$

Since, in level flight, L is fixed by the requirement that it equal the weight of the aircraft and ρ by the altitude, the only parameters left to use are C_L and S, both of which lead us to the use of high-lift devices and, in the case of C_L, an increase in alpha. The deployment of flaps or slats increases C_L and S, whereas increasing alpha increases C_L (Fig. 4-2).

Drag

An inevitable penalty of moving an object through a fluid is the production of drag, which is the resistance offered by the fluid to the motion of the body.

In very general terms, drag can be of four types:
- *Form drag* arises from the effort necessary to move the fluid away from the path of the object.
- *Viscous drag* results from the friction forces that exist between the moving body and the fluid.
- *Induced drag* is directly associated with the production of lift. A body that produces no lift has no induced drag associated with it.
- *Wave drag* arises in supersonic flight only. It is the drag associated with moving a shock wave through the fluid.

Ironically, the efficient operation of an aircraft might sometimes require that the drag actually be increased; such is the case when an aircraft is required to meet a speed and altitude restriction during descent from cruise altitude. The unbridled descent invariably will result in high speeds and a path that will not necessarily meet the restrictions imposed by ATC. In such cases, it is necessary to increase the drag of the aircraft by deploying spoilers to reduce the speed, increase the rate of descent, or both.

Another use for spoilers is to assist ailerons in rolling the aircraft at lower speeds. Indeed, some aircraft lack ailerons and depend exclusively on spoilers to produce roll, the advantage being that spoilers provide better adverse-yaw characteristics than ailerons.

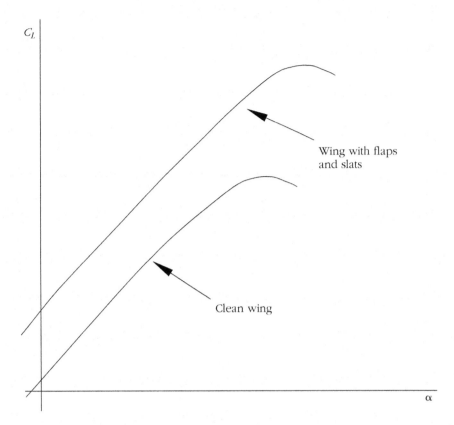

Fig. 4-2 *Effect of flaps on lift coefficient.*

5

Propulsion

A considerable portion of the sophistication of the modern jet transport can be found in the powerplant. The modern turbofan is endowed with fast and very capable computers, allowing an engine to operate at peak efficiency most of the time. Equally important are the advances introduced in metallurgy and high-temperature aerodynamics.

Again, all the benefits introduced by the higher technologies must be complemented by an equally sophisticated operation lest the operator fail to realize the most benefit in exchange for the considerable investment.

Turbojets and turbofans

All modern jet transports are powered by turbofan engines. The difference between the turbofan engine and its predecessor, the turbojet engine, can be discerned best by reference to Figs. 5-1A and 5-1B. The turbofan can be simply described as a turbojet with an additional stage (maybe two) of compressor machinery. The additional and most forward compressor stage feeds the *core* of the engine and also bypasses a substantial amount of air around the engine through the *bypass duct*. This *bypass air* is neither mixed with fuel nor burned. The benefits of such an arrangement are increased efficiency and lower noise generation.

The amount of air bypassed around the core divided by the volume of air fed through the core is known as the *bypass ratio* of the engine. Modern engines exhibit bypass ratios of 5-to-8 at the design flight condition.

Engine stations

Figure 5-1A shows also the stations that might be used to designate various locations in jet engines. These stations provide a ready reference to identify the points where the different propulsion parameters are measured; for example, $p_{2.5}$ refers to the static pressure at the face of the first compressor face.

Different types of jet engines have different station locations due to internal configurations that do not match station for station.

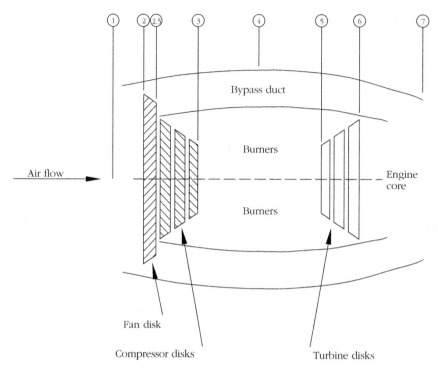

Fig. 5-1A *Schematic of a turbofan engine showing engine stations.*

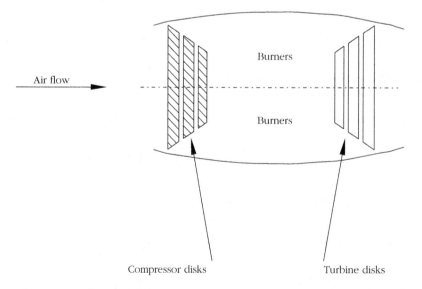

Fig. 5-1B *Schematic of a turbojet engine.*

Thrust-setting parameters (TSP)

The absence of a solid platform against which to react the thrust produced by the engine makes the direct measurement of thrust almost impossible while the aircraft is in flight. The need to know thrust accurately is nevertheless essential in the assessment of the performance of an aircraft in flight. To bridge the gap between these two realities, engine manufacturers resort to *thrust-setting parameters* (TSP).

A TSP is a quantity that can be measured on the ground and in flight and can be used to correlate conditions in both regimes of operation. During engine development, this quantity is measured and mapped against a large array of environmental conditions and thrust levels; thus, a given value of TSP is associated with a known set of conditions and level of thrust. A TSP allows us to know indirectly the thrust that the engine is producing by correlation with the thrust it is known to produce under the prevailing atmospheric and dynamic conditions.

Engine pressure ratio (EPR)

A common TSP used throughout the industry is *engine pressure ratio (EPR)*, which is defined as the ratio of the total pressure at the face of the low-pressure compressor and the total pressure at turbine exit. From Fig. 5-1, it is evident that this ratio is representative of thrust because it depends on the level of pressure remaining in the air flow after passing all the turbomachinery after all the power has been extracted by the turbines and other devices.

Figure 5-2A shows a typical plot of thrust versus speed for a given altitude (F_n is commonly used to denote thrust in the United States; it stands for force - net). Generally, as the altitude increases, thrust decreases for a constant value of EPR.

Alternatively, thrust can be presented as shown in Fig. 5-2B; it is then termed *corrected net thrust, F_{nc}*; it is the result of dividing F_n by δ, which yields a quantity that is less susceptible to altitude variations. Such a presentation is known as the *generalized thrust plot.*

Fan rotor speed (N_1)

Other engine manufacturers prefer to use N_1, the *fan rotor rotational speed*, as the parameter upon which to base thrust, with 100 percent N_1 being some arbitrary maximum determined by the engine manufacturer. Rarely will the user be exposed to an actual value of RPM; rather the manufacturer will have mapped all the pertinent engine performance parameters as a function of a percentage of N_1 ($\%N_1$).

This is not to suggest that at a rotational speed of 100 percent N_1 some structural limitation is encountered. The label of 100 percent N_1 is arbitrarily established by the manufacturer to simplify the performance presentation and provide a selected limitation to the operation of the turbomachinery.

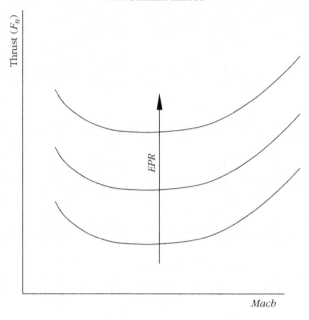

Fig. 5-2A *Net thrust versus Mach number and EPR at constant altitude.*

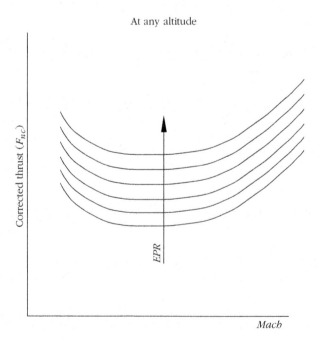

Fig. 5-2B *Corrected net thrust versus Mach number and EPR.*

Because most of the thrust produced by a turbofan comes from the bypass air, it is reasonable to expect that the rotational speed of the fan is a good indication of the thrust being produced.

Manufacturers will also use the rotational speeds of other rotors in the engine, $\%N_2$ and, when applicable, $\%N_3$, to characterize the overall performance of the powerplant regardless of the TSP used. Nevertheless, when rotor speed is used to characterize thrust, N_1 is the parameter used.

Often the rotor speed will be presented in corrected form to minimize the effects of altitude and speed. In such cases, N_1 is corrected and presented as

$$\frac{N_1}{\sqrt{\theta_T}}$$

Fuel flow (W_f)

Fuel flow, or fuel burn, is of great concern when dealing with efficiency. As with other engine parameters, W_f can be corrected, or its dependence on altitude minimized, through division by a selected atmospheric parameter. In the case of W_f, the denominator is the combination $\delta_{t2}\theta_{t2}{}^x$, where the exponent "$x$" is a characteristic of each engine.

The resulting parameter is the corrected fuel flow, W_{fc},

$$W_{fc} = \frac{W_f}{\delta_t \, (\theta_t)^x}$$

In some presentations, the fuel flow is not given directly; instead, fuel consumption is expressed in terms of the *specific fuel consumption* (SFC),

$$SFC = \frac{fuel\ flow}{thrust}\sqrt{\theta}$$

which is generally tabulated as $SFC/\sqrt{\theta}$ vs. F_{nc}. SFC is a measure of the amount of fuel flow required to produce a certain level of thrust, a measure that is appropriate because the fuel consumption of a jet engine is directly proportional (or very nearly so) to the thrust being produced. The correction $\sqrt{\theta}$ attempts to remove the effects of temperature and altitude variation.

As stated earlier: In general, W_f varies almost linearly with thrust, but also depends on environmental and flight conditions. It is presented in tables or plots such as the one shown in Fig. 5-3.

Thrust ratings

Thrust rating is the maximum level of thrust that an engine can attain under a given set of environmental and flight conditions. The concept is used to standardize comparisons between engines and to establish the limits to which an engine can be taken under different flight conditions.

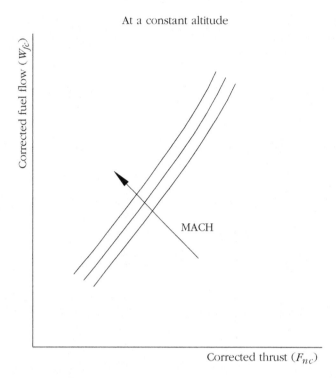

At a constant altitude

Corrected fuel flow (W_{fc})

MACH

Corrected thrust (F_{nc})

Fig. 5-3 *Corrected fuel flow versus corrected net thrust and Mach number.*

The following ratings are commonly used:

- *Takeoff* (TO)—the maximum thrust that can be extracted from an engine during the takeoff phase. It is usually limited to a duration of no more than 5 minutes.
- *Maximum climb* (MCL)—the maximum thrust that can be extracted from an engine during the climb phase.
- *Maximum continuous* (MCT)—the maximum thrust that can be extracted from an engine on a continuous basis. It can be used for an unlimited duration, and it is usually reserved for operations with a failed engine.
- *Maximum cruise* (MCR)—the maximum thrust that can be extracted from an engine while in cruise flight. It is usually reserved for altitude changes or speed increases during cruise.

A thrust rating is not a fixed value of thrust; rather, it is a set of values that depends on other pertinent parameters such as speed, altitude, and temperature. For example, as the aircraft climbs at an MCL rating, the actual thrust value, as measured in pounds of pushing force, will vary gradually as the speed of the airplane changes, as well as its altitude.

Engine manufacturers base the warranties they offer to their customers on these ratings. Indeed, the ratings are designed to limit the operations of engines such that normal wear and performance degradation of the machinery is not a source of dissatisfaction to the operators. The engine manufacturer and the operator will often enter into agreements by which the manufacturer guarantees the maintenance of the engine in exchange for reduced levels of rating operation; for example, if the operator commits to performing an agreed portion of all takeoffs at less than maximum power, the manufacturer will guarantee a minimum of maintenance over an agreed period of time.

Figure 5-4 contains a plot of a typical rating, in this case a takeoff rating. The horizontal segment of the curve is called the *flat rating* for obvious reasons. In theory, the variation of thrust with temperature follows a negatively sloped curve as shown in Fig. 5-4; but, at temperatures below where the flat segment begins (59°F in this case), the mass of air sucked into the engine contributes to the generation of pressure levels within the engine that violate the structural safeguards of the structure. The manufacturer, therefore, restricts the operation of the engine to thrust levels below this limit, calling it the *flat-rated thrust.*

As the ambient temperature increases beyond the flat-rating temperature (59°F for this example), the energy added to the air stream in the form of fuel increases the temperature of the gases reaching the turbine to values that might cause structural failure. Consequently, at the higher temperatures, the thrust level allowed must be gradually decreased to preclude thermal overstress. In ad-

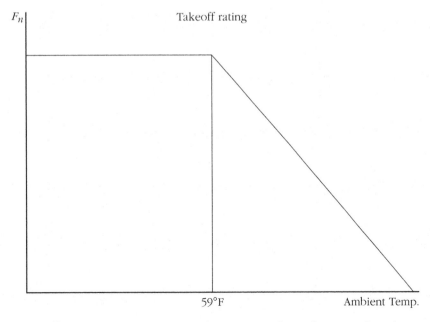

Fig. 5-4 *Takeoff rating expressed in terms of net thrust and ambient temperature.*

dition, increased temperatures result in less dense air and the associated decrease in the mass of air sucked into the engine. Both of these factors contribute to the negatively sloping line to the right of the 59°F point.

Although the idle setting in an engine is a recognized mode of operation, idle is not considered a rating. It is the minimum power setting used while decelerating or descending. The idle setting is designed to prevent engine flame-outs while keeping the minimum thrust level possible and as such will vary depending on the atmospheric and flight conditions that prevail.

Thrust limitations

The main thrust limitation imposed on a turbine engine is rotational speed. In the case of turbofans, this is expressed more commonly as N_1, the fan rotor speed. Other rotors might be present in the engine, but a limitation on the fan rotor carries with it an inherent limitation on the other rotors because their operation is interrelated by the internal dynamics of the engine as a whole.

Figure 5-5 shows the N_1 limitation at sea level as a line with negative slope. This line represents a constant value of rotational velocity: 5,000 RPM, for ex-

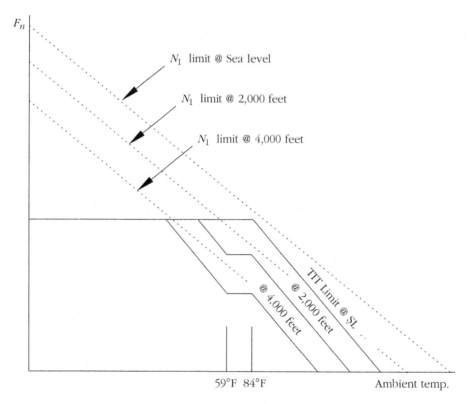

Fig. 5-5 *Thrust and* N_1 *limits for a typical thrust rating.*

ample. At this rotational speed, the rotor (fan, shaft, and turbines) experiences stress levels that will promote structural failure. These stresses are a result of centrifugal as well as aerodynamic loads. At the same value of rotor speed and a higher altitude, the engine will only produce a fraction of the sea-level thrust due to the decreased density of the air flow; therefore, the thrust associated with the limit rotational speed diminishes with altitude.

The pressure and turbine inlet temperature (TIT) limitations discussed previously are plotted on the same figure for SL and above. Note that for altitudes higher than SL there is an additional step on the curve before continuing to the flat-rated thrust level. This jog in the curve is designed to maintain the engine away from the corresponding N_1 limit line.

The plots shown in Figs. 5-4 and 5-5 are not the usual presentation offered to the operator by the engine manufacturer. A more common presentation is as shown in Fig. 5-6 within which EPR is used to indicate the level of thrust allowed under the particular rating. But because EPR will vary with altitude for the same level of thrust, the different EPR limits corresponding to each altitude are now specified. The concept of thrust limitation remains the same, but is now interpreted in terms of EPR instead of pure thrust.

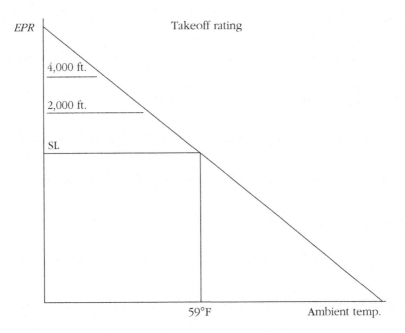

Fig. 5-6 *A thrust rating presented in terms of EPR limits.*

6

General performance topics

This chapter presents additional concepts, terminology, and tools that are useful in the study of aircraft performance. It will often mention the performance documentation provided by the manufacturer, who is the authorized source of numerous aircraft documents, some of which are dedicated to a comprehensive presentation of aerodynamic and propulsion data. These documents are a valuable tool in performance work.

Thrust available

Chapter 5 considered the notion of thrust in the context purely of propulsion, not linking it to the requirements of flight or its effect on flight dynamics. This thrust is known as *thrust available*, which is the thrust that the powerplant will deliver for a given throttle setting and the prevailing flight and atmospheric conditions (Fig. 6-1).

It is possible to present thrust available as a function of TSP, as shown in Fig. 6-2A, underscoring the fact that a TSP setting does not always translate to the same level of thrust.

Figure 6-2A is a representation of the effect of EPR on thrust available, for a specified altitude. To generalize the plot further, we can make use of corrected thrust, F_{nc}, and draw one plot that would apply for all altitudes, as shown in Fig. 6-2B. This presentation is commonly known as the *generalized thrust plot*, often found in the aircraft performance manuals.

Thrust required

Figure 6-3A shows an aircraft in equilibrium level flight. Under this condition, thrust equals drag, and lift equals weight. The thrust required is then exactly equal to the drag produced by the airframe: It is the thrust required to maintain level flight.

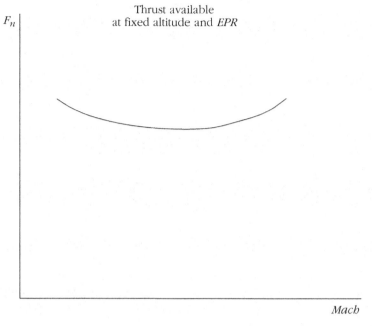

Fig. 6-1 *Thrust available versus Mach at a fixed altitude and EPR.*

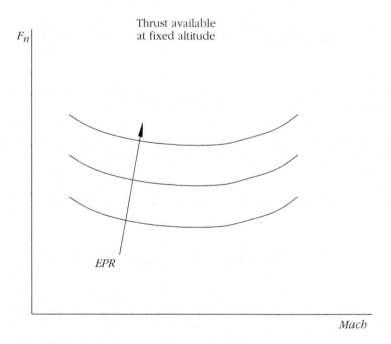

Fig. 6-2A *Dependence of thrust available on EPR.*

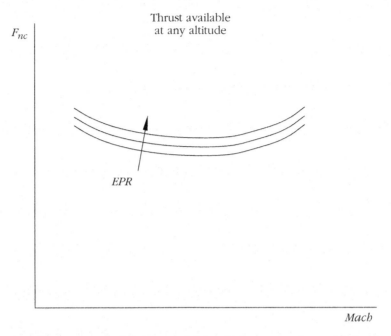

Fig. 6-2B *Corrected thrust available, also known as the generalized thrust plot.*

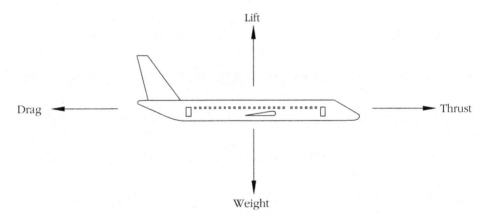

Fig. 6-3A *Forces acting on an airplane in equilibrium level flight.*

If the same aircraft were to be subjected to an increase in speed, an acceleration, the thrust required would then have to equal the drag plus the force required to accelerate the aircraft, as shown in Fig. 6-3B.

Thrust required is the term used to identify the amount of force that must be generated by the engines to overcome other forces, such as drag, inertia, and weight. It is therefore incorrect to assume that thrust required always equals drag.

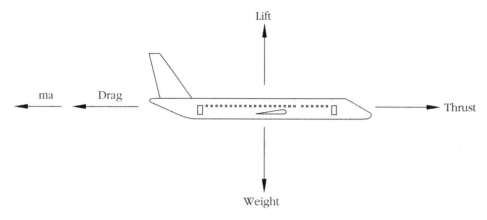

Fig. 6-3B *Forces acting on an airplane under level acceleration.*

In the analysis of performance problems, there is often the need to establish how much thrust is needed to execute a particular maneuver, so as to calculate the associated TSP and fuel flow that the crew will monitor on the flight deck. This is particularly true in modern aircraft, which are equipped with computers capable of calculating the parameters associated with the flight path to be flown and then relaying this information to the autopilot for hands-off operation.

Both thrust available and thrust required can be presented on the same plot, as shown in Fig. 6-4.

Bank-angle influence on G-loads (load factor)

Refer to Fig. 6-5 depicting an aircraft in a steady level turn; if we sum the forces along the vertical axis, parallel to the weight vector, the resulting equation is

$$\Sigma F = 0 \Rightarrow L\cos\phi = W \tag{6-3}$$

Under these conditions the lift will exceed the weight by the factor n:

$$L = nW \tag{6-4}$$

Substituting Equation 6-4 into Equation 6-3 yields Equation 6-5.

$$nW\cos\phi = W \Rightarrow n = \frac{1}{\cos\phi} \tag{6-5}$$

Through this relationship, the bank angle ϕ can be related to the g loading, n. We often see that ϕ and n share the same scale in a plot because they can be related through Equation 6-5.

At a fixed value of W/δ

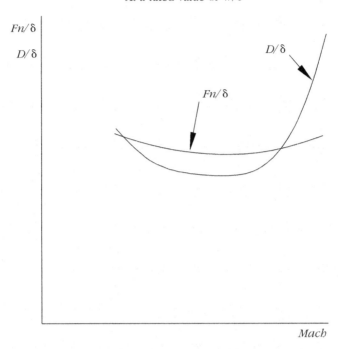

Fn/δ

D/δ

D/δ

Fn/δ

Mach

Fig. 6-4 *Variation of corrected thrust available and corrected thrust required with Mach number for a fixed altitude.*

Example 6-1

If an aircraft is limited to a 2-g load factor, what is the maximum bank angle that the aircraft can sustain in a level turn? From Equation 6-5,

$$2 = \frac{1}{\cos\phi} \Rightarrow \cos\phi = 0.5 \Rightarrow \phi = 60°$$

Buffet limits

Two forms of aerodynamically induced vibrations are *buffet* and *flutter*. The latter is the object of considerable attention by the designers of the airframe because of its destructive nature and consequently intolerable presence in an aircraft.

Whereas flutter is an aeroelastic phenomenon capable of destroying the aircraft in flight, buffet is a relatively harmless vibration that by itself presents no danger to the airframe. Buffet, nevertheless, might be the harbinger of a worse condition, namely a *stall*.

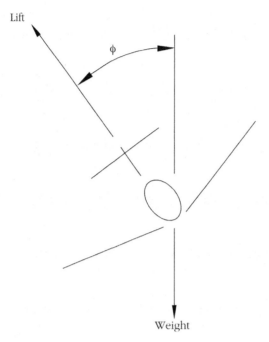

Lift

φ

Weight

Fig. 6-5 *Forces acting on an airplane in an unaccelerated level turn.*

As the speed is reduced and α is increased, the flow past the wing is forced to go over an increasingly more difficult obstacle. At some point, the flow begins to separate, first intermittently (thereby causing buffet), and finally permanently, resulting in the complete loss of lift (the stall) and perhaps also loss of control (Figs. 3-4 and 3-5). Here buffet serves to signal the approaching stall. Where buffet is not apparent because of the particular design of the wing, an artificial *stick shaker* is introduced to alert the crew to the impending stall.

At the higher speeds (Fig. 3-6), small shock waves form over the wings; these in turn trigger unsteady conditions behind them, thereby causing buffet. Again, by itself, this buffet is harmless, but it does precede a condition where total flow separation might occur and with it loss of control due to the disturbed flow over the aerodynamic control surfaces.

Aircraft manufacturers thus establish buffet boundaries to designate the regime of flight beyond which flight should not be attempted lest buffet be encountered. These limits are usually presented in plots such as Fig. 6-6.

From Equation 3-6B we can solve for the ratio of aircraft weight to local pressure ratio, W/δ, and plot the curves of constant W/δ as shown in Fig. 6-6.

$$\frac{W}{\delta} = (1481.68)\ C_L M^2 S \qquad (6\text{-}6)$$

The benefit of using the ratio as opposed to only using weight is that the plot is then valid for any altitude.

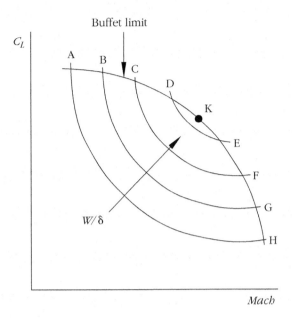

Fig. 6-6 *The buffet limit curve.*

Points A, B, C, and D of Fig. 6-6 represent conditions of slow-speed stall. Points E, F, G, and J represent conditions of high-speed stall. Point K is informally known as *coffin corner* for a good reason: An aircraft at point K is very near the point where any action will result in a stall. Bear in mind that any change in speed will result in a displacement along the constant W/δ line. At point K, any such displacements will result in an aggravation of the situation. The only alternative left to the nervous crew in this situation is to do absolutely nothing and wait until the fuel burned changes the weight of the aircraft, thereby shifting its W/δ curve down, where the gap between low- and high-speed buffet widens. Rewrite now Equation 3-6 as:

$$\delta = \frac{W}{(1481.681)\,C_L\,M^2 S} \tag{6-7}$$

Select a value of M and the corresponding C_L from Fig. 6-6. Then, selecting a W will yield a δ from Equation 6-7, which in turn will yield an altitude from the standard atmospheric tables or Equations 1-4 and 1-5 once solved for h. It is possible then to plot altitude versus Mach number for initial buffet onset at a g loading of unity, as shown in Fig. 6-7.

We also can interpret Fig. 6-6 slightly differently; instead of having various lines of constant W/δ, we can call these lines of constant multiples of one value of W/δ. The plot now would have lines of constant nW/δ, where n is the number of g forces to which the aircraft is subjected. For example, in Fig. 6-8, the lowest constant-W/δ line is for a value of 200,000 lb. The other constant-W/δ line is for a value of 400,000 lb., or twice the original value. For a constant alti-

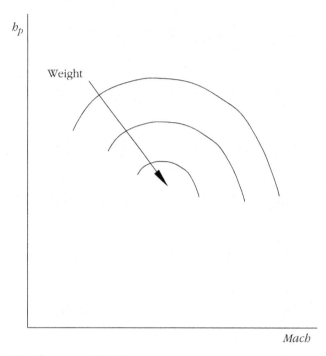

Fig. 6-7 *Initial buffet onset.*

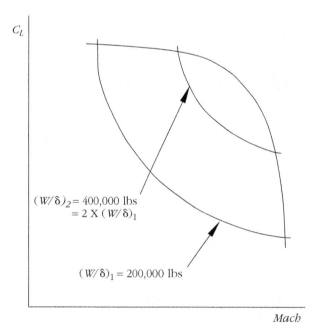

Fig. 6-8 *Example of buffet limit curve.*

tude and δ, we might choose to interpret this as an increase of 200,000 lb. in weight or an increase of twice the g force to which the airframe is subjected; either way, the wing has to generate twice the lift, and, in so doing, shift closer to the buffet limit curve.

It is possible then to plot points A through J from Fig. 6-6 against M in terms of an nW/δ versus M plot, where the value of W/δ at A would be the basic value and all others are multiples of it. The resulting plot, shown in Fig. 6-9, is useful in determining the ability of an aircraft to sustain buffet-free g loads (pull ups, banked turns, etc.) at a specific value of Mach number.

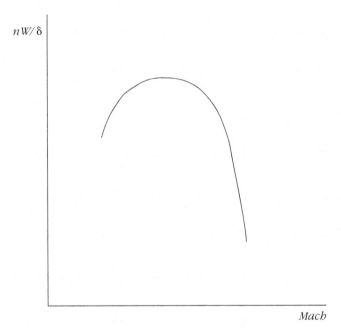

Fig. 6-9 *G-load capability as a function of Mach.*

Example 6-2

The aircraft of the accompanying figure is in cruise at FL 300, weighing 180,000 lb. (A) What is the maximum Mach No. under these conditions? (B) If the cruise altitude is FL410, what is the minimum Mach for these conditions? (C) If we arbitrarily establish that $W/\delta = 1,400,000$ is the absolute limit beyond which the aircraft cannot be flown, what is the maximum bank angle allowed if the aircraft weighs 250,000 lb. and cruises at FL370?

(A) We need to find the W/δ corresponding to the given condition:

$$\frac{W}{\delta} = \frac{180,000}{\delta_{FL300}} = \frac{180,000}{0.2970} \approx 600,000 \ lb.$$

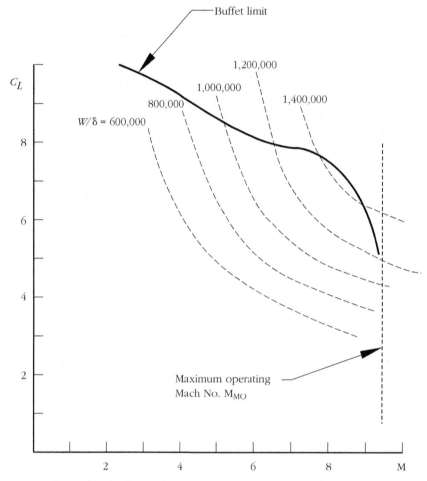

Figure for example 6-2.

Following the line for W/δ = 600,000 lb., we see that it meets the M_{MO} boundary before meeting the buffet boundary for the aircraft; therefore, the maximum Mach allowed is the *maximum operating mach*, M_{MO}, or approximately 0.94.

(B) Following a similar procedure as in (a), first we calculate W/δ:

$$\frac{W}{\delta} = \frac{180,000}{\delta_{FL410}} = \frac{180,000}{0.1764} \approx 1,020,000 \; lb.$$

Again following the line of constant W/δ at a value of 1,020,000 lb., we find that the minimum Mach is approximately 0.56.

(C) First we need to establish the value of W/δ that applies at the condition of interest:

$$\frac{W}{\delta} = \frac{250,000}{\delta_{FL370}} = \frac{250,000}{0.2138} \approx 1,170,000 \; lb.$$

Then we need to calculate how much larger the absolute maximum W/δ is than the value just calculated (the ratio n), and from that, with Equation 6-5, determine the maximum bank angle:

$$\frac{\left(\dfrac{W}{\delta}\right)_{max}}{\left(\dfrac{W}{\delta}\right)_{FL\,370}} = \frac{1,400,000}{1,170,000} = 1.1966 = \frac{1}{\cos\phi} \Rightarrow \phi \approx 33.3°$$

Also with the value of W/δ for FL370, we can see that in this regime of flight the Mach number must be between the limits of approximately 0.62 and 0.94 to avoid initial buffet.

Bank-angle influence on turn radius

Figure 6-10 depicts an airplane in a level, constant speed turn. From vector mechanics, we know that the centrifugal acceleration is the quotient of the linear velocity squared and the turn radius; the centrifugal force is then the product of this ratio and the aircraft mass.

Summing the forces along the horizontal axis

$$\Sigma F_b = \frac{V^2}{R} \cdot \frac{W}{g} - L\sin\phi = 0 \qquad (6\text{-}8)$$

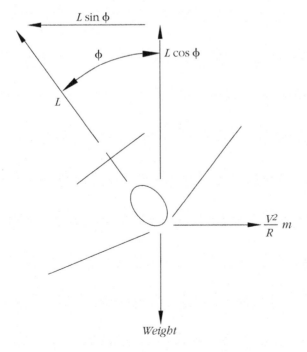

Fig. 6-10
Forces acting on airplane in a constant-speed turn.

Summing now the forces along the vertical axis

$$\Sigma F_v = L \cos\phi - W = 0 \qquad (6\text{-}9)$$

Combining both expressions in a division of $\sin\phi$ by $\cos\phi$ yields the relationship we seek between turn radius and bank angle,

$$\tan\phi = \frac{V^2}{R \cdot g} \qquad (6\text{-}10\text{A})$$

or

$$R = \frac{V^2}{g \cdot \tan\phi} \qquad (6\text{-}10\text{B})$$

Example 6-3

An aircraft climbs at $V_2 = 140$ KCAS after suffering an engine failure during take-off from an airport situated in a valley. At an altitude of 5,000' MSL, the procedure calls for the pilot to initiate a left turn, away from the highest of the mountains, back to the airport. In the direction of the turn, but no closer than 4 nautical miles (nm), there is another mountain before which the turn must be completed. Will a 15° bank be sufficient to complete the turn safely?

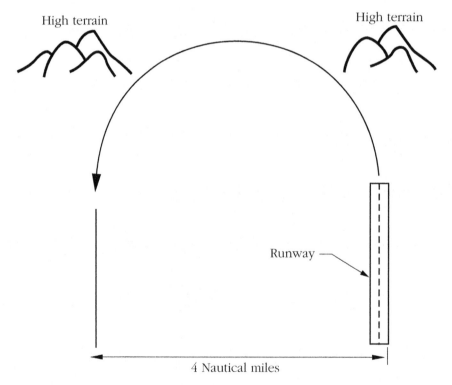

Figure for example 6-3.

To solve this problem, we must first convert CAS to TAS, and to do this, we first convert CAS to Mach using Equation 2.5:

For alt. = 5,000', δ = 0.8321; substituting these values into Equation 2.5 yields M = 0.232.

At 5,000' the velocity of sound is 650 knots; therefore, TAS = 650 × 0.232 = 151 KTAS, which also equals 255 fps.

Substituting these values into Equation 6-10B:

$$R = \frac{(255)^2}{32.2 \tan(15°)} = 7548'$$

and this equals 1.24 nm, indicating that the turn diameter is shorter than the distance to hill.

If the operator furthermore decided that it is a good idea to pad the procedure by assuming that the pilot can be counted to hold 10°, thereby enlarging the turn, the resulting turn radius would be

$$R = \frac{(255)^2}{32.2 \tan(10°)} = 11,452'$$

The answer is 1.89 nm, which still results in a turn diameter within the prescribed 4 nm, but with a much reduced margin.

The reader will note that although the aircraft is climbing, we have used the actual TAS to calculate the turning radius and bank angle. Strictly speaking, the TAS, even in the absence of wind, does not equal the ground speed during a climb, but the angle of climb subtended by an aircraft climbing with a failed engine is small enough to be neglected. In such small-angle problems, it is customary to simplify the problem by assuming that

$$\cos\phi = 1$$
$$\sin\phi = 0$$

If our problem had been to accommodate an aircraft climbing during normal operation, the assumption of a small climb angle would not be valid, as can be ascertained by anyone who has watched the normal takeoff of a modern jetliner. In such a case, the ground speed, which determines the dynamics of the turn would be calculated as:

$$ground\ speed = (TAS)\cos\phi$$

In the presence of wind, of course, the ground speed equals the TAS minus the wind velocity.

Bank-angle influence on climb gradient

Figure 6-11A shows an aircraft climbing in a turn at a constant TAS. Figure 6-11B is a rendition of the same aircraft viewed along its longitudinal axis. The need

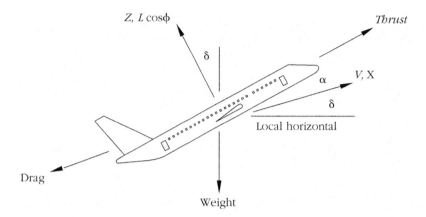

Fig. 6-11A *Forces acting on an airplane in a climbing turn, view along lateral axis.*

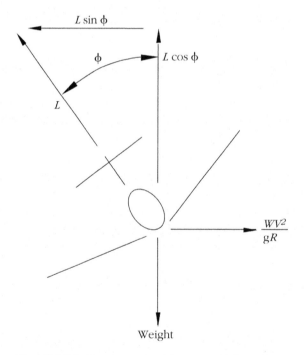

Fig. 6-11B *Forces acting on an airplane in a climbing turn, view along longitudinal axis.*

to use two figures is evident when considering that the lift vector is inclined by both the angle of bank ϕ and the climb angle γ. This subsection explains how ϕ affects γ.

First, from Fig. 6-11B, it is evident that the component of L along the z-axis is only $L\cos\phi$. Then, summing the forces along the z-axis in Fig. 6-11A,

$$\Sigma F_z = 0 \Rightarrow L \cdot \cos\phi = W \cdot \cos\gamma \qquad (6\text{-}11)$$

and along the x-axis

$$\Sigma F_x = 0 \Rightarrow T \cdot \cos\alpha = W \cdot \sin\gamma + D \qquad (6\text{-}12)$$

Solving for sin γ and cos γ in the above expressions and dividing one by the other yields

$$\tan\gamma = \frac{T \cdot \cos\alpha - D}{L \cdot \cos\phi} \qquad (6\text{-}13)$$

Equation 6-13 illustrates the relationship that exists between γ and ϕ. To generate an actual trajectory point by point, it is necessary to solve simultaneously Equations 8-1 and 8-2 of chapter 8, a task that is rarely necessary in normal airline operations, but nevertheless occasionally encountered.

More commonly, the operator finds it is necessary to assess the impact of gradient degradation due to a banked turn. For this, aircraft manufacturers provide the operator (usually in the flight manual) an accurate assessment of the amount of γ degradation that the aircraft will suffer as a result of banking. This information is presented in terms of a gradient decrement as a function of bank angle and flap setting, thereby alluding to its importance in takeoff path analysis.

These data are of paramount importance in performing the obstacle avoidance calculations mandated by the regulatory agencies.

To understand the use of the information provided in the estimation of obstacle clearances, consider Fig. 6-12, where the gradient without bank is

$$\frac{y + \Delta y}{x} = \frac{y}{x} + \frac{\Delta y}{x} \qquad (6\text{-}20)$$

In other words, the aircraft is capable of climbing $y+\Delta y$ feet as it covers x feet of ground. If the altitude to which the aircraft can climb is reduced by Δy due to the banked attitude that it has during a turn, the climb gradient will be reduced by $\Delta y/x$, a quantity that is provided by the manufacturer.

One method of solving gradient degradation problems consists of increasing the height of the obstacle to be overflown by an amount that corresponds to the gradient degradation given. Instead of degrading the gradient in the turn for the purpose of calculating obstacle clearance, we raise the height of the obstacle by an equivalent amount and assume an unbanked nondegraded straight path.

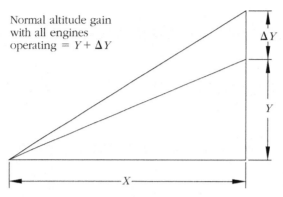

Fig. 6-12 *Definition of gradient degradation parameters.*

To accomplish this, the obstacle height is raised by an amount Δy,

$$\Delta y = x \cdot \left(\frac{\Delta Y}{x}\right) \tag{6-21}$$

The distance x is the distance traversed in banked flight only since this is the only portion of the trajectory where the gradient is degraded.

Example 6-4

The performance engineers at XYZ-Air have devised a path for an aircraft to follow in the case of an engine failure shortly after taking off from the runway shown, which is located at an altitude of 5,000 feet MSL. They know that the hill in the heading of the runway must be avoided, but are not too sure if the 2,000-foot hill under the large turn can be cleared. They know the aircraft will be flying at a speed of 150 KCAS at point 1, at flap setting 10, and 200 KCAS with zero flaps at point 4. The procedure calls for the first turn, at point 2, to change the heading by 60°, and the 180° turn at point 4 to begin abeam point 5. All turns are to be at a bank angle of 15°. Finally, the gradient degradation is a function of the flap setting as shown below in percent;

Bank angle	Flaps = 0	Flaps 10
5	0.02	0.04
10	0.10	0.20
15	0.30	0.65

By how much should the engineers raise the height of the obstacle at point 5 to correctly account for the degradation in gradient?

First we translate all the speeds to units we can use in our formulas:

150 KCAS @ 5,000' $\Rightarrow$ M=0.248 $\Rightarrow$ 161 KTAS $\Rightarrow$ 272 fps
200 KCAS @ 5,000" $\Rightarrow$ M=0.331 $\Rightarrow$ 215 KTAS $\Rightarrow$ 363 fps

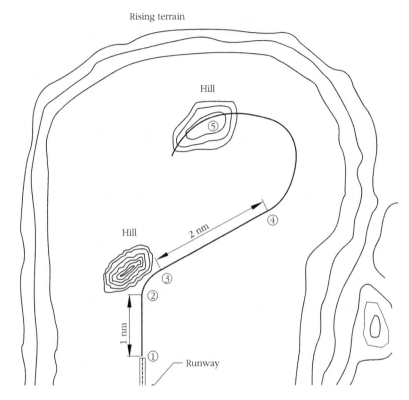

Figure for example 6-4.

Now we calculate the radius of turn at each turn point:

$$R_2 = \frac{V^2}{g\,\tan\phi} = \frac{272^2}{32.2\,\tan15°} = 8574'$$

$$R_4 = \frac{V^2}{g\,\tan\phi} = \frac{363^2}{32.2\,\tan15°} = 15{,}272'$$

With the radii, the arc distance covered while in a bank can be calculated:

$$s_{2-3} = R\theta = 60\left(\frac{\pi}{180}\right)8574 = 8979'$$

$$s_{4-5} = R\theta = 15{,}272\pi = 47{,}979'$$

Finally, the incremental height due to each turn can be calculated:

$$\Delta h_{2-3} = \left(\frac{0.65}{100}\right)8979 = 58.4'$$

$$\Delta h_{4-5} = \left(\frac{0.30}{100}\right)47{,}979 = 143.9'$$

The total increment in the height of the obstacle is then 143.9 feet plus 58.4 feet, for a total of 202 feet.

Radius of turn
in the presence of winds

Often it is necessary to determine the ground track of an aircraft flying inside a moving mass of air. When this mass of air moves directly with or against the aircraft motion, the problem is simplified to the tailwind or headwind problem respectively. In such cases the ground velocity is:

$$V_g = TAS \pm V_W \tag{6-22}$$

More generally the problem is as presented in Fig. 6-13. Here an aircraft flies in a southeasterly direction, and at point "O" initiates a turn at a constant bank angle. The case of a variable bank angle can be accommodated by a piecewise application of the procedure presented here.

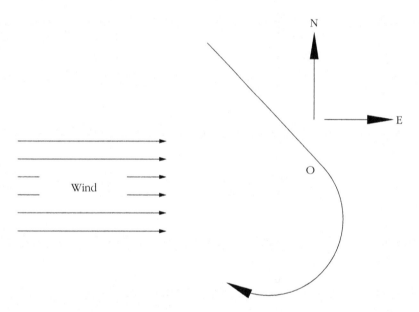

Fig. 6-13 *Schematic of a turn in the presence of wind.*

This kind of problem can be solved by using the relations of particle kinematics. We first establish arbitrarily a system of coordinates parallel to the local magnetic or geographic directions, as the case might suit. The origin of this set of coordinates will be at the point where the aircraft initiates the turn. It will be further assumed that the bank angle is available instantaneously, which, although not possible, incurs insignificant errors.

If there is no wind, a constant bank angle will result in a circle being traced both in the air and as a projection on the ground, which is the *ground track*.

Figure 6-14 shows a more detailed presentation of the variables of interest:

- θ_0 is the angle subtended between the radius of turn and the local north direction. It is equal to the heading angle minus 90° for positive rotations, or heading angle plus 90° for negative rotations. A path that increases the heading angle as it progresses has a positive velocity associated with it and is said to describe a positive rotation; θ_i and θ_{i+1} are subsequent values of the angle θ.
- θ_w is the angle between the wind direction and the local north direction. Notice that this is strictly a geometrical presentation and that winds are traditionally reported in terms of the direction from which the wind blows, not in terms of the direction toward which it blows, as required by this presentation; the difference is exactly 180°.
- r is the radius of turn of the aircraft in the air mass; it will equal the radius of the ground track only when there is no wind.
- V_w is the wind velocity vector.
- V is the aircraft velocity vector. The magnitude of this vector is positive if the aircraft turns in a direction that increases its heading; it is negative otherwise.
- i is a subscript indicating a particular point along the path.

Earlier we saw that the turn radius can be expressed as a function of the bank angle and flight speed, the relationship being

$$r = \frac{V^2}{g\tan\phi} \tag{6-23}$$

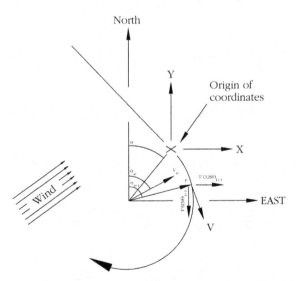

Fig. 6-14 *Turning in the presence of wind with parameter definitions.*

This expression will be necessary in the following development to calculate r.

The motion around the circular track can be divided into two separate motions, one along the y-axis, and one along the x-axis. Beginning at the onset of the curved path, the first x coordinate on the curved track is determined from the expression

$$x_1 = x_0 + V\cos(\theta_1)\,(t_1 - t_0) \tag{6-24A}$$

where both x_0 and t_0 are zero, since the turn starts at the origin, and we arbitrarily start time at zero. Since $V\cos(\theta_1)$ is the velocity component along the x axis, multiplying by the quantity $(t_1 - t_0)$ yields the distance traveled along in the x direction during the time span $(t_1 - t_0)$.

Likewise, along the y axis,

$$y_1 = y_0 - V\sin(\theta_1)\,(t_1 - t_0) \tag{6-24B}$$

where both y_0 and t_0 are also zero.

More generally, for all points along the curve, the expressions are

$$x_i = x_{i-1} + V\cos(\theta_i)\,(t_i - t_{i-1}) \tag{6-25A}$$

$$y_i = y_{i-1} - V\sin(\theta_i)\,(t_i - t_{i-1}) \tag{6-25B}$$

The angle θ_i is made up of two components: the original angle θ_0 due to the initial heading, and the angle accrued by virtue of the rotational motion, which can be expressed as

$$\theta_i = \theta_{i-1} + \frac{Vt_i}{r} \tag{6-26}$$

The coordinates of each successive point along the circular path can now be written as:

$$x_i = x_{i-1} + V(t_i - t_{i-1})\cos\left(\theta_0 + \frac{V_{t_i}}{r}\right) \tag{6-27A}$$

$$y_i = y_{i-1} - V(t_i - t_{i-1})\sin\left(\theta_0 + \frac{V_{t_i}}{r}\right) \tag{6-27B}$$

In the presence of wind, the above expressions are augmented by one term each:

$$x_i = x_{i-1} + V(t_i - t_{i-1})\cos\left(\theta_0 + \frac{Vt_i}{r}\right) + V_w\,(t_i - t_{i-1})\sin\theta_w \tag{6-28A}$$

$$y_i = y_{i-1} - V(t_i - t_{i-1})\sin\left(\theta_0 + \frac{Vt_i}{r}\right) + V_w\,(t_i - t_{i-1})\cos\theta_w \tag{6-28B}$$

This development finds much use in the study of special departure procedures where its implementation in a spreadsheet or coded in a simple program yields valuable results in assessing the effect of various alternatives in obstacle avoidance.

Example 6-5

An airplane flying at 179 KTAS, in 164° heading and in the presence of a wind of 254° at 20 knots, begins a turn to the left with a 15° bank angle. Estimate when it will reach a heading of 270°.

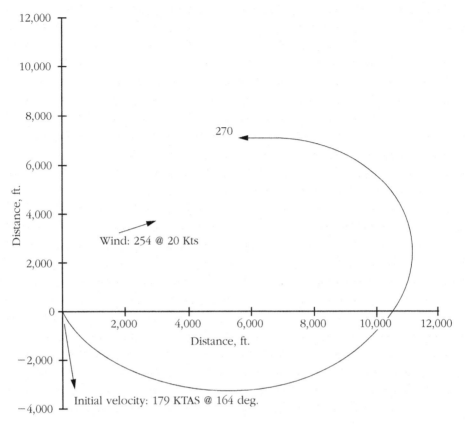

Figure for example 6-5.

Based on the bank angle and speed, the radius of turn can be calculated using Equation 6-10b,

$$R = \frac{V^2}{g \tan 15^\circ} = \frac{302^2}{32.2 \tan 15^\circ} = 4917'$$

Note that the speed has been used in units of feet per second. The initial input parameters are then:

- Wind magnitude = 20 knots, which equals 34 fps
- Wind direction = from 254°, or, to 74°
- Cos (Wind D) = 0.28
- Sin (Wind D) = 0.96
- Bank angle = 15°
- Aircraft speed = −179 KTAS, or −302 fps
- Radius of turn = 4,917 ft
- Initial angle = 164°

Then, using Equation 6-28 in a spreadsheet, along with the other information, results in the output shown in Table 6-1 and its plot.

Table 6-1 Table for example 6-5

Heading (°)	t (sec.)	X (ft.)	Y (ft.)
164.00	0	0	0
146.38	5	999	−1,212
128.77	10	2,340	−2,112
111.15	15	3,912	−2,611
93.54	20	5,584	−2,658
75.92	25	7,212	−2,244
58.30	30	8,661	−1,403
40.69	35	9,809	−210
23.07	40	10,563	1,227
5.46	45	10,869	2,779
−12.16	50	10,713	4,303
−29.78	55	10,125	5,662
−47.39	60	9,175	6,731
−65.01	65	7,967	7,417
−82.62	70	6,630	7,657
−86.15	71	6,361	7,687
−89.67	72	6,091	7,698

As shown in the table, the aircraft reaches the target heading after about 72 seconds, some 6,091 feet east and 7,698 feet north of where the turn began. The heading shown at the end is −89° because the arithmetic followed here does not recognize the boundary at 0°/360° and continues the process of subtraction.

V-n diagrams

The *V-n* diagram derives its name from the variables plotted along its axes: velocity (V) and load factor (n). This diagram is generally used to depict both the maneuvering capability of the aircraft and its ability to fly in gusty conditions.

The boundaries of the envelope, as shown in Fig. 6-15, are defined as follows:

- 1 to 2. The constant load factor represented by this line is selected by the manufacturer, but cannot be less than 2.5.
- 2 to 3. This boundary is established by the diving speed, V_D.
- 3 to 4. This line is defined by the linear variation between the negative load factor at the design cruise speed, V_C and V_D.
- 4 to 5. As the positive load factor, the negative value is selected by the manufacturer, but cannot be greater than −1.
- 0 to 1. This line is defined by a plot of Equation 6-29.

$$n = \frac{C_{Lmax}\left(\frac{1}{2}\rho S\right)V_{sn}^2}{W} \tag{6-29}$$

(Equation 6-29 is a variation of Equation 3-6(A), where *n* is the load factor attained at the maximum lift coefficient, C_{Lmax}, and flying at the stall speed, V_{sn}, for that load factor.)

- 0 to 5. This line has a similar definition to the 0-to-1 line, but using the maximum negative C_L.

The following parameters are also represented in Fig. 6-15:

- V_{S1}. The stall speed at a load factor of one, which, as expected, is found along the C_{Lmax} curve.
- V_C. The design cruise speed chosen by the manufacturer.
- V_A. The maneuvering speed, defined by the FARs as

$$V_A = V_{s1}\sqrt{n}$$

where *n* is the maximum allowable load factor (2.5 for commercial aircraft).

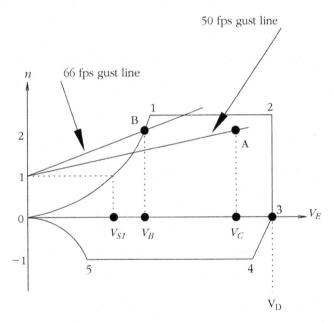

Fig. 6-15 *The* V-n *diagram.*

In addition, Fig. 6-15 shows the velocity V_B, the design speed for flight in maximum gust intensity, and whose definition warrants further detail:

The FARs define the maximum gust intensity in the cruise condition as a 50 fps vertical wind to be encountered at the design cruise speed; this is shown as point A in Fig. 6-15. The 50-fps line serves to define n_{gust} at the cruise speed V_C. Likewise, the 66-fps line is used to define point B, whose significance will be evident in a subsequent paragraph of this chapter.

Figure 6-16 shows the velocity vectors for the cruise speed and gust speed, which in turn define the incremental α to which the aircraft is subjected when it suddenly encounters a gust. Furthermore, under the assumption that V_G is much smaller than V_C, it is valid to approximate $\Delta\alpha$ as V_G/V_C. The change in C_L due to this change in α is then

$$\Delta C_L = C_{L_\alpha} \Delta\alpha = C_{L_\alpha} \frac{V_G}{V_C} \tag{6-30}$$

where C_{L_α} is the rate of change of C_L with respect to angle of attack, α.

In turn, the change in total lift due to this change in C_L is

$$\Delta L = q S C_{L_\alpha} \frac{V_G}{V_C} = \frac{1}{2}\rho V_C S C_{L_\alpha} V_G \tag{6-31}$$

The change in load factor arising from the increment in lift is

$$\Delta n = \frac{\Delta L}{W} = \frac{\rho C_{L_\alpha} V_C V_G}{2\dfrac{W}{S}} \tag{6-32}$$

Finally, the total load factor due to the encountered gust is

$$n = 1 + \frac{\rho C_{L_\alpha} V_C V_G}{2\dfrac{W}{S}} \tag{6-33}$$

The resulting value of n applies to a theoretical sharp-edge gust and does not include the effects of the aircraft's inertia. In practice, these two effects are taken into account by multiplying n by statistical factors that reduce its value. Note that the load factor is directly proportional to both the cruise and gust speeds and inversely proportional to the wing loading.

Fixing the gust speed at 50 fps at the design cruise speed, V_C, defines n for a specified set of cruise conditions. This ordered pair of cruise speed and the resulting n defines point A in Fig. 6-15, and the line joining this point to $n = 1$, on

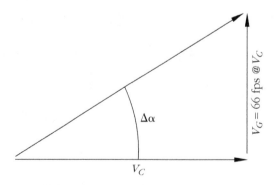

Fig. 6-16 *The geometry of cruise speed and gust speed vectors.*

the *n*-axis, is termed the "50-fps gust line." At speeds lower than V_C on this line, the values of V_G and forward speed will be different from those associated with point A but will maintain a linear relationship among themselves.

Similarly, B is defined by the intersection of the 66-fps gust line and the 0-to-1 line in Fig. 6-15; V_B is its corresponding speed. A lower limit of V_B is

$$V_B = V_{s1}\sqrt{n_{gust}@V_C}$$

Additionally, various other gust lines (−25 fps, 25 fps, 30 fps, etc.) can be derived following a similar reasoning. Indeed, these lines are used to establish different requirements of the flight regime. For more details on the regulatory requirements, see FAR Parts 25.335 and 25.341.

Maximum speeds

From the previous definitions of thrust available and thrust required, we can conclude that the intersections of the two curves (Fig. 6-4) represent limiting flight conditions. At these intersections, the thrust available equals the thrust required; therefore, the aircraft is in an equilibrium condition where neither the speed nor the rate of climb can change without a shift in either the drag or the thrust curve.

More specifically, if a limiting thrust condition is chosen, such as *maximum cruise rating* (MCR), the line representing this condition will intersect several D/δ lines that correspond to different values of W/δ (Fig. 6-17). These curves are intersected twice by the MCR curve, a circumstance that is not necessary but certainly possible. Since the plot was made for one altitude, each of points 1 through 8 has associated with it a value of weight and *M*. Plotting these points for a fixed altitude results in Fig. 6-18, where we have included also points A through J of Fig. 6-6, the buffet boundary. The distance separating the buffet boundary curve from the curves of Fig. 6-18 gives an indication of the margin that exists between the current condition at MCR thrust (or whatever thrust level was chosen for Fig. 6-17) in level flight and the onset of buffet—all of which can assist in determining the maneuver capability of the aircraft at that condition.

For example, if $(W/\delta)_{buff}$ is associated with the onset of buffet and $(W/\delta)_{FL350}$ corresponds to FL350 while at MCR, their ratio will be:

$$\frac{\left(\dfrac{W}{\delta}\right)_{buffet}}{\left(\dfrac{W}{\delta}\right)_{FL350}} = n$$

From Equation 6-5, the maximum bank angle ϕ possible before entering a buffet condition is

$$\phi = \arccos\left(\frac{1}{n}\right)$$

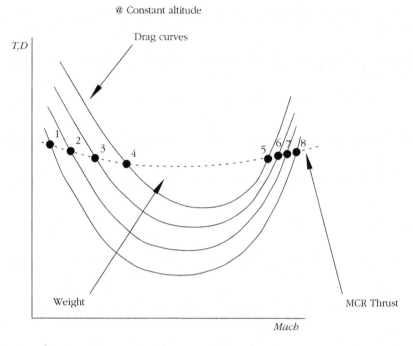

Fig. 6-17 *Intersection of thrust and drag lines.*

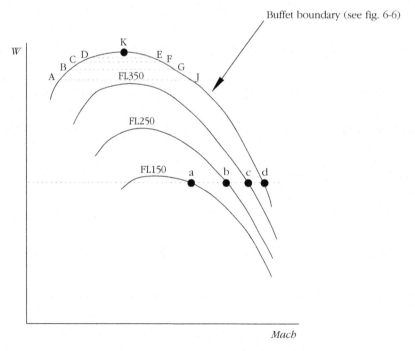

Fig. 6-18 *Buffet margin.*

7

Takeoff

Regardless of the capability of an aircraft to operate efficiently in flight, it is the takeoff maneuver that will determine how much weight can be carried aloft. Little is gained from an aircraft that is very fuel efficient if, during the takeoff, the payload lifted is not enough to at least pay for the trip undertaken. In this chapter, the reader will be exposed to some of the more salient reasons why a modern jet transport cannot operate from just any runway and why the savvy operators will always manipulate all the variables at their disposal to maximize takeoff capability.

Takeoff is the maneuver required to take the aircraft from a static condition on the runway to either:

- An altitude of 1,500 feet above the field elevation, or
- An altitude at which transition to the enroute configuration is completed.

In the United States, the takeoff process is regulated by the Federal Aviation Regulations (FARs); this requires that the aircraft be certified under Part 25 and that the operator be certified to operate under either Parts 119, 121, or 135. Other countries have similar regulations, but the presentation of the material here follows the requirements imposed by the FARs.

The regulations do not specify any of the speeds, distance, or takeoff weight (TOW) that are to be produced by the takeoff calculation; but the regs do specify the relationships that exist among the different speeds and between the various distances involved in the takeoff maneuver. It is then the responsibility of the operator to ensure that, in the quest for a profitable operation, the limitations imposed are observed.

It might be thought that in assessing a takeoff, the operator would proceed by calculating the runway necessary to lift a desired aircraft weight at the prevailing atmospheric and runway conditions. Unfortunately, this simple conception of the takeoff problem leads to unwieldy calculations. For example, an operator might select a *takeoff weight* (TOW) for a given departure for which it would then be necessary to calculate aircraft acceleration and deceleration rates, maximum velocities attainable given a fixed runway length, liftoff point, climb gradient, and the like. In the event that sufficient velocity for continuing a takeoff with an engine failure is not attainable or not enough runway remained to abort the takeoff safely, an iterative process would begin, reducing the TOW until all parameters met the legal requirements.

Airlines are in the business of air transport and cannot dedicate resources to such tedious calculations. With this in mind, the manufacturers have prepared the results of enormous amounts of similar computations and charted them in the AFM, allowing the operator to complete the process in reverse order. Instead of iterating through several trials in an attempt to find the TOW allowable, the operator can instead establish the prevailing conditions (runway length, wind, temp, runway slope, etc.) and look up the corresponding allowable TOW.

The material presented in this chapter is intended to help in understanding the conditions and limitations imposed on the process of computing the data that the operator finds in the AFM, and eventually also understanding the use of the AFM in optimizing the operation of the aircraft. What follows is not a guide for operators to do their own computations; such a guide would have to be very particular to the aircraft being addressed.

It is pertinent to point out at this time that the AFM, in its paper form (usually a three-ring binder, some 4 inches thick), is rarely used in computing anything for the regular operation of a fleet of aircraft. All aircraft manufacturers have computerized AFMs, resulting in complex programs capable of producing results in the quantities and speed required by most modern operators. Indeed, some manufacturers are beginning to supply the AFM only in digital form for ready use in a computer.

In an effort to simplify the presentation, we will conceive the takeoff as consisting of two portions, each having its own environment:

- The earthbound portion, which is limited by the conditions of the runway
- The airborne portion, which is limited by the environment around the airfield and the performance of which the aircraft is capable

The link between these two portions of the takeoff maneuver is the speed V_1, defined below.

Takeoff speeds

Before the discussion can continue on the subject of takeoff, some speed definitions are necessary (Fig. 7-1):

- "V_{mcg}, the minimum control speed on the ground, is the speed during the takeoff run at which, when the critical engine is suddenly made inoperative, it is possible to recover control of the airplane with the use of primary aerodynamic controls alone to enable the takeoff to be safely continued. . . ." (FAR Part 25.149)
- V_{ef} is the airspeed at which the critical engine is assumed to become inoperative.
- V_1 is the airspeed beyond which no stopping action will be initiated by the flight crew. This speed is commonly referred to as the *decision speed*: a nomenclature with which this author disagrees because no

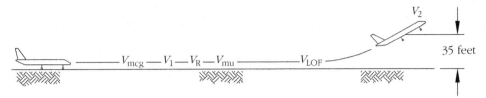

Fig. 7-1 *Selected takeoff-speed definitions.*

decision is made at this speed. Seconds before reaching V_1, in the event of an engine failure, the captain will recognize the failure (failure recognition speed, V_{ef}) and then make the decision to abort or not depending on whether it is possible to initiate the abort process before reaching V_1.

- V_{mbe} is the maximum airspeed on the ground from which a stop can be accomplished within the energy-absorbing capacity of the brakes.
- V_{mca} (or V_{mc}) "is the airspeed at which, when the critical engine is suddenly made inoperative, it is possible to recover control of the airplane with that engine still inoperative and maintain straight flight with either zero yaw or with a bank angle of no more than 5°." (FAR Part 25.149)
- V_R is the airspeed at which rotation is initiated during takeoff to obtain V_2 (subsequently defined in this list) at 35 feet above the liftoff point.
- "V_{mu}, the minimum unstick speed, is the airspeed at which the airplane can be made to lift off the ground and to continue the takeoff without displaying any hazardous characteristics." (FAR, Part 25.107d)
- V_{LOF} is the airspeed at which the airplane becomes airborne.
- V_{te} is the airspeed beyond which the rotational speed of the tires is excessive.
- V_s is the stall speed of the airplane in the takeoff configuration (FAR Part 25.103). V_s is important because it constitutes the foundation for the definition of most of the takeoff speeds.
- V_2 is the target speed at the 35-foot height and the speed that must be maintained to achieve the required climb gradients in the airborne portion of the takeoff.

The definitions above have been listed in the order in which they would be observed during takeoff (Fig. 7-2).

Among these speeds there are some constraints:

- $V_{mcg} \leq V_1 \leq V_R$
- $V_1 \leq V_{mbe}$
- $1.05\,V_{mca} \leq V_R$
- $1.1\,V_{mu} \leq V_{LOF}$, all engine takeoff
- $1.05\,V_{mu} \leq V_{LOF}$, one engine inoperative takeoff
- $V_{LOF} \leq V_{te}$
- $1.1\,V_{mca} \leq V_2$
- $1.2\,V_s \leq V_2$

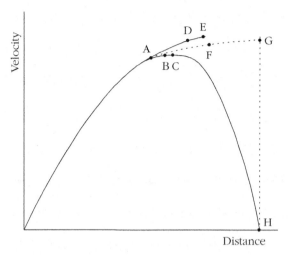

Fig. 7-2 *Sequence of speeds along takeoff process.*

Runway configuration

The requirement that the aircraft must either be brought to a stop or successfully taken airborne within the confines of the available runway and its extensions places great emphasis on the distance available to execute these maneuvers. The following definitions will be necessary for our discussion:

- A *clearway* is an area beyond the runway, not less than 500 feet wide, centrally located about the extended centerline of the runway and under the control of the airport authorities (FAR Part 25).

- A *stopway* is an area beyond the runway, no less wide than the runway and centrally located about the extended centerline of the runway. It must be able to support the airplane during an aborted takeoff without causing structural damage to the airplane (FAR Part 25).

- The *accelerate-stop distance* (ASD) is the distance required to accelerate the aircraft to V_{ef} with all engines operating, experience an engine failure, accelerate to V_1 with one engine inoperative, and finally reduce the speed to zero.

- The *takeoff distance* (TOD) (Fig. 7-3) is the greater of the horizontal distances from the start of takeoff to the point at which the airplane is 35 feet above the takeoff surface with one engine inoperative or 115 percent of the horizontal distance from the start of takeoff to the point at which the airplane is 35 feet above the takeoff surface with all engines operating.

- The *takeoff run* (TOR) (Fig. 7-4) is the greater of the horizontal distance from the start of takeoff to a point equidistant between the point where the liftoff speed is reached and the point where 35 feet is attained with

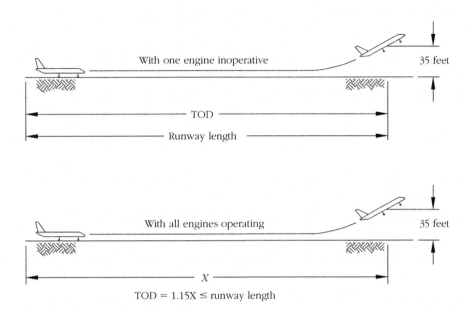

Fig. 7-3 *Takeoff distance (TOD).*

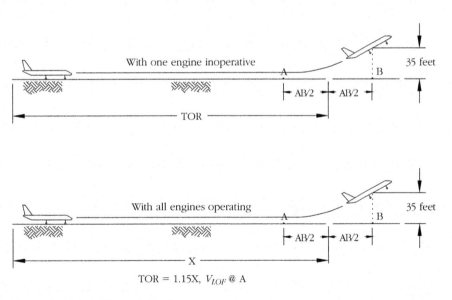

Fig. 7-4 *Takeoff run (TOR).*

one engine inoperative or 115 percent of the horizontal distance from the start of takeoff to a point equidistant between the point where the liftoff speed is reached and the point where 35 feet is attained with all engines operating.

The *takeoff field length* required by the FAR can be defined now as the distance that allows the least TOW of the following:

- In the case of a *rejected takeoff* (RTO), the distance required to bring the aircraft to V_{ef} with all engines operating, recognize the engine failure, initiate stopping action before V_1, and then bring the aircraft to a stop.
- In the case of a continued takeoff with an engine failure, the distance required to accelerate to V_{ef} with all engines operating and then to continue the takeoff such that a height of 35 feet is reached.
- With all engines operating, 115 percent of the distance required to accelerate, lift off, and reach a height of 35 feet above the runway.

To reiterate: The allowable takeoff weight of the aircraft is the least of the weights allowed under each of the three conditions listed above.

The earthbound portion of the takeoff

To become airborne, the airplane must gain enough speed to produce the required lift; unfortunately, the process of accelerating to the required speed takes place in a limited length of runway where, if something were to go wrong, the crew has little time to assess the situation.

Under the FARs, all takeoffs must be calculated and executed under the assumption that the most critical engine becomes inoperative at a point along the takeoff run. This point is defined in terms of V_1. Below V_1, the takeoff must be safely aborted; above V_1, the takeoff must be continued safely and the problem resolved in the air. To study this eventuality, it is helpful to refer to Fig. 7-2, where we have depicted the sequence of possible events associated with an engine failure and the speeds at which these occur.

From the start of the takeoff run to point A, the airplane accelerates normally with all its engines.

At point A, the airplane suffers an engine failure, and its thrust decays rapidly but gradually. The residual thrust of the engine as it spools down must be considered in calculating acceleration.

At point B, the crew has recognized the engine failure and initiated stopping action by retarding the throttles.

At point C, the brakes are applied and the spoilers deployed.

At point H, the aircraft comes to a complete stop.

Point F is related to the situation in which the engine failure occurred beyond A, thus making it impossible to stop the aircraft in the available stopping distance; in this case the takeoff must be continued, resulting in the aircraft becoming airborne at F, powered only by the thrust provided by the remaining engines.

At G, the aircraft, with one engine inoperative, reaches the takeoff safety speed V_2 at a height of 35 feet.

If no engine failure occurs, D is the point at which the aircraft reaches liftoff speed.

At point E, the aircraft, powered by all engines, reaches the takeoff safety speed V_2.

The condition depicted in Fig. 7-2 is known as a *balanced field condition* (BFC), where the distance required to stop the aircraft after an engine failure exactly equals the distance traveled to reach the 35-foot height at V_2, also with a failed engine. A BFC is not a required condition; it is merely a peculiar condition in which the ASD equals the TOD.

The main parameters that affect the condition of a field and the capability of the aircraft to take off from it are: altitude, runway length, engine thrust, wind, runway slope, aircraft weight, and ambient temperature. The set of conditions that results in a BFC in the morning might not be there in the afternoon because the wind and temperature might have changed. We see then that the length of the field is only one of the parameters that influence the BFC.

It is also important to recognize that not all fields are balanced, nor do they have to be. It is entirely possible and acceptable that a runway be such that the distance covered in stopping the aircraft in an aborted takeoff be more, or less, than the distance traveled to reach the 35-foot height. The only stipulation is that neither of these distances can exceed the length of the available runway and its extensions, the stopway and clearway, which are subsequently examined in this chapter.

The rejected takeoff (RTO)

There are arguably more situations that call for an aborted takeoff than those that warrant the continuation of a takeoff after an engine failure. Should anything happen to the aircraft before reaching V_1 that the captain deems detrimental to the continued takeoff, the takeoff must be aborted, and, by the very conditions that determine V_1, the aircraft will come to a stop within the ASD.

In practice, neither the decision to abort nor the outcome of the decision are as neat. In fact, only two decisions are associated with an unexpected event during the takeoff roll: whether the event is detrimental to flight safety and whether V_1 has been reached; therefore, V_1 is not, as is commonly and mistakenly thought, a decision speed in the sense that a decision must be made at that speed—the pertinent decisions have been made already. V_1 is a decision speed only in the sense that it constitutes a boundary beyond which no decision can be made, and the takeoff must be continued into flight.

The preceding statement is a reflection of the regulations governing the takeoff process. There are situations in which some flexibility must be allowed. For example, it is not clear that the takeoff could be safely continued if an engine were to experience a catastrophic failure and cause substantial damage to the wing. Aborting the takeoff, even at speeds higher than V_1, might be a better policy in this case. It is in cases such as this that nothing can be a substitute for the experience of the crew.

Figure 7-5 will facilitate the understanding of the mechanics of an RTO, where the most critical engine in the aircraft is assumed to fail. In a simple runway, without the benefit of a stopway or a clearway, the ASD available equals the runway length (Fig. 7-5A).

By adding a stopway, as shown in Fig. 7-5B, we have increased the space available to bring the aircraft to a complete stop without increasing the runway length.

By increasing the ASD, a stopway violates the definition of a BFC, but it might allow an increased TOW. Furthermore, a longer ASD implies a higher V_1 than would apply for a BFC because now there is more distance available to bring the aircraft to a stop. Careful study of Figs. 7-5A and 7-5B reveals the importance of V_1 in the dynamics of the takeoff maneuver. If the takeoff is to be aborted after an engine failure, a lower V_1 will result in a shorter ASD. It takes a relatively shorter distance to accelerate to a lower V_1; therefore, it takes a shorter distance to stop the aircraft from this lower V_1.

If we now add a clearway but no stopway, the resulting condition is shown in Fig. 7-5C.

The clearway does not contribute to the stopping effort, but it does provide more room to achieve the 35-foot height once the aircraft becomes airborne. The clearway thus adds more room to accomplish the takeoff, much as the stopway contributed to the stopping effort, but in both cases neither addition contributed to both efforts.

Consider now the condition resulting from adding both a clearway and a stopway, as shown in Fig. 7-5D. The case is the same as in Fig. 7-5B; the clearway contributes nothing to the stopping effort. It is possible, though, to have a clearway that meets the requirements of a stopway, as can be discerned from the definitions of stopway and clearway.

Stopways and clearways can be seminal to the profitable operation of an aircraft. Either one or both might mean the difference between operating an aircraft with enough fuel to reach a distant destination with a profitable payload or alternatively reducing either the range or the payload to levels that impair profitability.

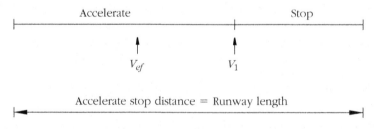

Fig. 7-5A *Rejected takeoff on a simple runway.*

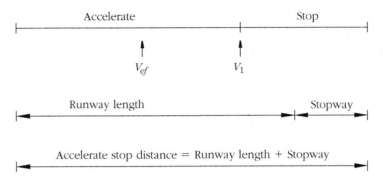

Fig. 7-5B *Rejected takeoff with a stopway.*

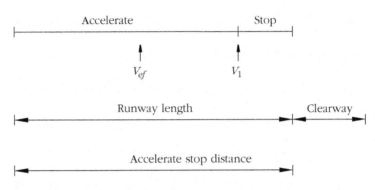

Fig. 7-5C *Rejected takeoff with a clearway.*

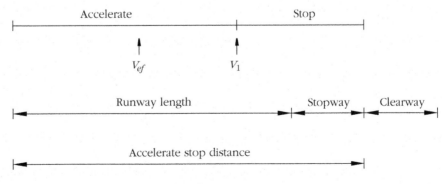

Fig. 7-5D *Rejected takeoff with stopway and clearway.*

The continued takeoff

There are two conditions under which the takeoff process is continued in pursuit of flight: that, in the event of an engine failure, no stopping action has been initiated before V_1 and also that no engine failure occurs. The latter is expected in normal everyday operations.

Unnatural as it might seem to the pilot, continuing the takeoff after reaching a speed equal or higher than V_1 produces a successful liftoff and climb to 35 feet before reaching the end of the runway and possible clearway, even if the effort involved is considerable and subsequent climb gradient shallow.

Figure 7-3 shows that the TOD must not be greater than the runway length. It is also conceivable and admissible that, under the conditions of this takeoff (weight, wind, runway slope, temperature, etc.), if the crew had elected to abort, the aircraft would come to a stop at some distance before the end of the runway. In other words:

$$ASD < AGD = RUNWAY\ LENGTH$$

where the accelerate-go distance (AGD) equals the available runway length, but the ASD is less—there is no BFC here.

Another alternative situation is that

$$ASD > AGD = RUNWAY\ LENGTH$$

which is not allowed because it would mean that the aircraft stops beyond the end of the runway if the takeoff were to be aborted.

The benefit in pointing out these two possibilities lies in the illustration of a third possibility. The takeoff distance exactly equals the ASD, which, in turn, equals the runway length:

$$ASD = TOD = RUNWAY\ LENGTH$$

Again, this is known as a balanced field condition (BFC), or balanced field length. The latter terminology, although widely used, is sometimes confusing because the length of the field is merely one of the elements that defines a BFC.

Recall that according to the FARs, takeoffs must be predicated upon an engine failure during the takeoff. The BFC is noteworthy because it is the condition under which the maximum aircraft takeoff weight (TOW) can be realized from a specific runway (without clearway or stopway) under the prevailing atmospheric conditions. Under an unbalanced condition, the ASD and the TOD requirements are not the same, and the one that produces the least TOW prevails. Bear in mind, though, that the resulting TOW must be compared to the TOW resulting from requiring that 115 percent of the TOD fit within the runway; again, the least TOW prevails.

If a clearway were added to the available runway, the resulting condition appears in Fig. 7-6. Note that the addition of a stopway would not affect this case since the aircraft is assumed to continue the takeoff. In Figs. 7-5C and 7-5D, the clearway was added to underscore its ineffectiveness on the aborted takeoff. A similar argument can be followed here with the stopway and its uselessness in the continued takeoff.

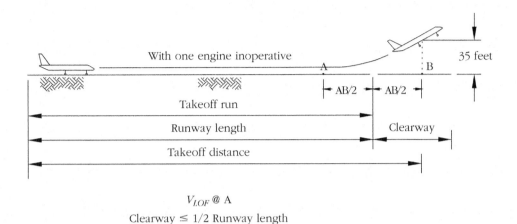

V_{LOF} @ A

Clearway ≤ 1/2 Runway length

Fig. 7-6 *Continued takeoff with a clearway (one engine inoperative).*

If the takeoff is not aborted following an engine failure, a high V_1 will require a shorter TOD since the aircraft has less speed to gain on the remaining engines to reach V_R and V_{LOF}. A low V_1 will require a longer AGD because the aircraft must accelerate over a wider speed range to reach V_R from the lower V_1.

Note that the addition of a clearway immediately negates a BFC. With the clearway, the available TOD is now longer than the ASD, thus violating the definition of a BFC, namely that ASD equal TOD.

The addition of a clearway might increase the TOW by allowing a longer TOD, which in turn allows a lower V_1 than the V_1 for a BFC. We now have that TOD > ASD.

Earlier we said that adding a stopway and increasing V_1 would result in an increased TOW. We are now saying that with the addition of a clearway and a reduction of V_1, a similar result can be obtained. Strange as it seems, both an increase and a decrease in V_1 might result in a higher TOW, depending on the field conditions. A high V_1 might be desirable to raise V_R and V_2 and thereby enhance the climb capability of the aircraft. A low V_1 is desirable when the stopping conditions are degraded, in which case the only other alternative is to reduce the TOW so that the aircraft mass can be stopped before reaching the end of the available stopping distance.

If, in the course of calculating the takeoff parameters for an aircraft, it is found that the limiting condition is the ASD, then increasing V_1 and adding a stopway will contribute to a profitable solution.

Alternatively, if the limiting condition is the TOD, then a reduction in V_1 and the addition of a clearway will be the solution sought. This solution is also applicable to the case in which the limitation comes from the all-engines-operating requirement, namely that 115 percent of the TOD be available on the runway.

Consider now the case of a takeoff with all engines operating; we will not consider a stopway since there is no requirement to evaluate the aborted takeoff when all engines are operating. Without a clearway, the condition is shown in Fig. 7-3.

With the addition of a clearway, the condition changes, as shown in Fig. 7-7; X_R is the distance from the beginning of the takeoff roll to a point halfway between the liftoff point and the point where the aircraft reaches a height of 35 feet, and X_D is the distance required for the aircraft to achieve a height of 35 feet.

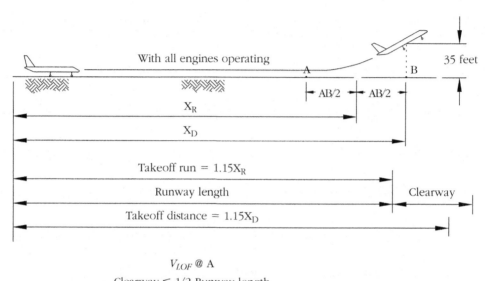

V_{LOF} @ A

Clearway $\leq$ 1/2 Runway length

Fig. 7-7 *Continued takeoff with a clearway (all engines operating).*

From the above discussion we can discern that V_1 does not affect the all-engine takeoff directly. The effect is felt indirectly through the notion that, under FAR Part 121, the takeoff must be planned assuming that the most critical engine will fail, and that either under normal all-engines-operating conditions, or in the event of an engine failure, the takeoff must not be aborted after reaching V_1.

By plotting ASD and TOD against V_1 for a fixed TOW (Fig. 7-8), we can gain some insight into their relationship. These graphs are to be interpreted as the ASD or TOD required to achieve the corresponding V_1 in the horizontal axis. They are not plots of the distance required to achieve that V_1; in other words, they are not plots of acceleration, but of the V_1 that will result in the corresponding ASD or TOD. Therefore, given the BFC at $V_{1,BAL}$, reducing V_1 to $V_{1,D}$ results in TOD_C and ASD_D. The distance required to accelerate to any of these values of V_1 is of course much shorter.

Let us suppose now that, for some reason, there is a requirement to increase V_1 to $V_{1,B}$. We already know that the balanced field length is represented by A; therefore, the only way that we can meet the requirement that the aircraft has enough stopping distance is to have enough runway or stopway such that the

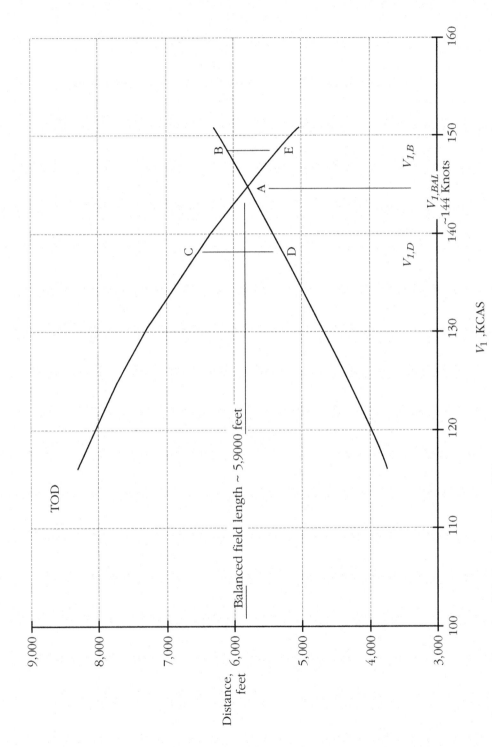

Fig. 7-8 *Variation of* TOD *and* ASD *with* V_1.

ASD available be represented by B. If the actual runway length is, for example, 7,000 feet, the higher V_1 can be accommodated. Note that for the higher $V_{1,B}$, the TOD requirement dropped to E.

Similarly, if for some reason, we prefer a lower V_1, say $V_{1,D}$, the associated requirement for TOD is now at C. To comply with this requirement, we need to have enough runway or clearway to bring the available TOD to the distance represented by C. Again, a 7,000-foot runway would accommodate the lower V_1. Note that for the lower V_1, the ASD requirement dropped to D.

Figure 7-8 applies to one TOW. If we want to increase the TOW of the aircraft, the curves would shift, as shown in Fig. 7-9 by the dashed lines; both ASD and TOD distance requirements have increased for the higher weight W_2. The 7,000-foot runway we had before can still accommodate this heavier takeoff, but with less margin to change V_1 as we did before at the lower weight. Note that the balanced field V_1 has increased from $V_{1,1}$ to $V_{1,2}$.

Multiple V_1

We have seen that V_1 is driven only by the dynamics of the takeoff roll; its value is not affected by what happens to the aircraft after it becomes airborne.

V_1 is limited at the low end of the scale by the controllability of the aircraft ($V_1 \geq V_{mcg}$) and at the higher values by aerodynamic requirements ($V_1 \leq V_R$) and the capability of the brake system of the aircraft ($V_1 \leq V_{mbe}$); therefore, V_1 can be selected as low as V_{mcg}, provided there is enough room to successfully complete a takeoff should an engine fail and the takeoff continued. V_1 can be as high as V_R or V_{mbe}, whichever is lower, provided there is enough room to stop the aircraft should the takeoff be aborted before reaching V_1.

A high or low V_1 is chosen within the available range based on other operational considerations, some of which are:

- Slippery or contaminated runways might dictate the operator's policy regarding the choice of V_1 under these conditions.
- Safety considerations and preference of the operator.
- The need for a higher V_2 and the enhanced climb gradient that it brings with it.

Figure 7-10 shows some of the effects of the variation of V_1 on ASD and TOD. Point 1 refers to the BFC (TOD = ASD = 6,000 feet, and V_{1BAL} is the V_1 corresponding to a BFC) that will allow the TOW to equal about 101,000 lb. If, due to other constraints or a light load, the weight is reduced to 97,000 lb. (represented by the short segment just below the 100,000-lb. line), an infinite choice of values for V_1 becomes available, namely all the points along the 97,000-lb. line that lie between points 2 and 3. Note that at 3 V_1 is very close to being limited by V_R; whereas at 2, V_1 is about 99 percent of V_{1BAL}.

At point 2, V_1 is lower than V_{1B}, and, as a consequence, the margin of safety is increased should the takeoff be aborted—the speed from which the aircraft must be stopped is slower and there is considerably less energy to be dissipated

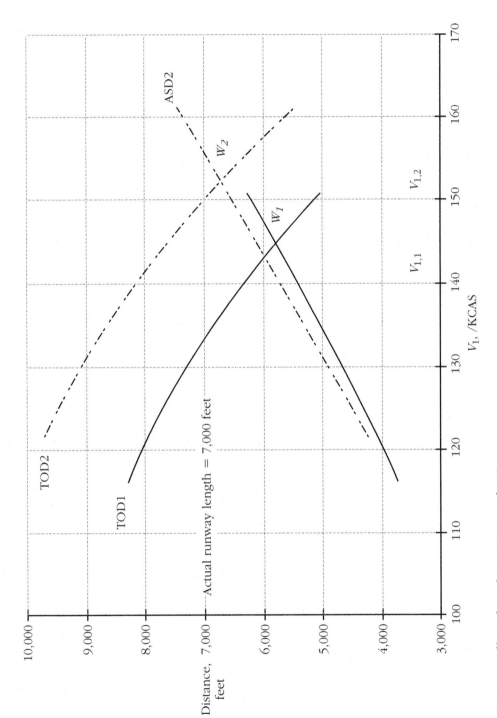

Fig. 7-9 *Effect of weight on* TOD *and* ASD.

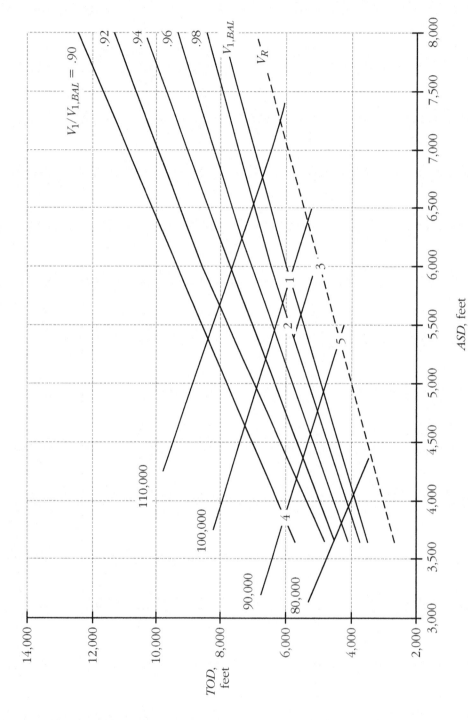

Fig. 7-10 *Effect of* V_1 *on* TOD *and* ASD.

by the brakes. The drawbacks of the slower V_1 are that it exactly meets the required clearance at the end of the runway for the case of a continued takeoff after an engine failure and that there is less margin with respect to minimum control speeds.

At point 3, the above situation is reversed: V_1 is higher than V_{1BAL}, resulting in a reduced TOD in the event of a continued takeoff after an engine failure, and with it comes an increased clearance at the end of the runway. Of course, since the speeds are higher, any concern of violating the minimum control speeds is reduced. The penalty paid for these advantages is that, in the event of an aborted takeoff, the aircraft will consume all 6,000 feet to stop. For all the intermediate points between points 2 and 3, the conditions will vary as a function of their location along the 97,000-lb. line.

If the TOW were to be reduced to 90,000 lb., the V_1 associated with point 4 might not be available, depending on the location of the V_{mcg} boundary. Certainly, point 5 is not a valid takeoff condition because its associated V_1 is greater than V_R.

From this discussion, we conclude that a range of V_1 will be available as long as there remains some margin in either the TOD or the ASD. Conversely, it is also true that a single V_1 exists when both the ASD and the TOD available have been completely used up.

Unbalanced field condition

Due to the influence of wind, runway slope, or some other parameter, it is possible that the required ASD will not equal the required TOD, in which case, by definition, the takeoff is executed under an unbalanced condition. Points 2, 3, and 4 of Fig. 7-10 represent such conditions. Point 2, for instance, might represent a takeoff on a downward sloping runway with no stopway, which affords an ASD somewhat shorter than d_1. On the other hand, the runway might have a substantial clearway, thus making TOD = d_1 > ASD. From Fig. 7-10, it is evident that such a set of conditions prescribes a unique V_1, namely the one corresponding to point 2.

A similar reasoning can be followed in establishing that point 3 is associated with the condition that TOD < d_1, while ASD = d_1.

Similarly to the previous section, it is evident that if the actual TOW is less than the TOW that would exhaust both the TOD and the ASD, a range of V_1 is also available. If, under the unbalanced condition represented by point 2, the actual TOW were W_2 and not W_3 (the maximum allowed at point 2), a range of V_1 would be available, namely all the V_1 values from point 4, along the W_2 line, to the point where ASD equals the ASD for point 2.

Example 7-1

The chart on Fig. 7-10 applies to a typical twin-engine transport. The table below contains information about the relationship between weight and takeoff speeds:

Table 7-1
Example 7-1

W	V_{1BAL}	V_R
100,000	130	133
110,000	136	140
120,000	145	151
130,000	152	160
140,000	160	170

The runway from which the takeoff is to be attempted is 6,000 feet long but has a stopway of 1,000 feet and a clearway of 500 feet. At 100,000 lb., the aircraft will complete the takeoff with all engines operating in 4,000 feet. Are the runway characteristics such that the takeoff can be made legally?

From the chart and the fact that the TOD = 6,000 + 500 = 6,500 feet, the corresponding V_1/V_{1BAL} for 100,000 is in the range from about 0.950 to 1.00; this corresponds to an ASD range of 5,300 to 5,900 feet respectively. The ASD range falls well within the available ASD of 7,000 feet (6,000 + 1,000), which fulfills that requirement.

The all-engine takeoff can be accomplished in 4,000 feet, to which we are required to add 15 percent to calculate the all-engine TOD:

$$4,000 \times 1.15 = 4,600 \text{ feet} < 6,500;$$

which also meets the requirement. But, V_1 can have a range:

$$V_1 = 0.950 \times 130 = 123$$

$$V_1 = 1.00 \times 130 = 130$$

Actually, V_1 can be increased up to V_R, for a reduction in TOD required to 4,700 feet and an increase in the ASD required to 6,400 feet.

Runway conditions

As presented in the aircraft flight manual (AFM), the takeoff performance of the aircraft is predicated on the use of a dry runway where the friction afforded by its surface is at a maximum. Unfortunately, runways are often wet, icy, or contaminated by snow or water. The effect of these contaminants on the runway is manifested in different ways, depending on the actual conditions and the particular aircraft.

Wet and icy runways make braking less effective, resulting in an extended stop portion of the ASD. Besides reducing the TOW, V_1 must also be reduced to accommodate the need for more stopping distance under slippery conditions. Conversely, a runway with an accumulation of snow or slush on its surface will inhibit the aircraft's ability to accelerate, thereby increasing the AGD.

Many methods and devices minimize the effects of moisture and contaminants on runway surfaces:

- Grooved runway pavement promotes fast drainage of water and provides a rougher surface for improved friction.
- Antiskid systems installed on aircraft inhibit a braking wheel's tendency to lock up.
- The removal of rubber deposits from the runway eliminates a source of friction breakdown.
- Aircraft manufacturers publish advisory data on the takeoff performance degradation due to runway contaminants. These may be used by the operator to tailor specific policies regarding operation on contaminated runways.

As of this writing, no regulatory agency has produced regulations dealing with the takeoff or landing performance on slippery or contaminated runways. Partially this is the result of a void in the understanding of the dynamics of an aircraft operating under these conditions, and, appropriately, the subject of much ongoing research.

The airborne portion of the takeoff

Once the pilots have successfully negotiated the acceleration of the aircraft to V_{LOF} with an engine inoperative, they must fly the aircraft in a path that meets the minimum requirements established by the FARs. Important issues pertinent to the ensuing flight path are:

- A minimum climb gradient must be maintained.
- The minimum climb gradient must be increased to overfly any obstacle under the flight path.
- Banked turns are limited by the impaired performance of the aircraft flying with an inoperative engine.
- Takeoff thrust is available for a limited period of time.
- The flight crew's workload is increased substantially after an engine failure.

We have seen that the ground portion of the takeoff introduced a number of requirements that resulted in limitations on the TOW. Similarly, the airborne portion of the takeoff is governed by regulations that also limit the TOW, which will be covered in the following material. The requirement—from either the ground or airborne portion of the takeoff—that yields the least TOW is the one that rules.

Takeoff segments

After the aircraft has successfully become airborne, the ensuing flight path must follow strict guidelines to ensure the safety of flight required by the regulations and good professional practice. The airborne part of takeoff is divided into segments, each of which is addressed specifically by the FARs, and whose general description is as follows:

First segment comprises the interval from the end of the TOD (the point at which the aircraft reaches 35 feet above the takeoff surface at V_2) to the point where the landing gear is fully retracted. During this portion of the takeoff, the aircraft is assumed to be at takeoff power and the flaps in the takeoff configuration (no flap retraction allowed). The gear is retracted during the first segment.

Second segment begins at the end of the first segment and extends to a gross height of at least 400 feet (called the *level-off height*), although some operators choose to extend the second segment to greater heights for operational reasons. While in the second segment, the aircraft is assumed to maintain takeoff power and flap setting.

Third segment begins at the end of the second segment. While in this segment, the flaps are retracted, the power may be reduced, and level-flight acceleration is undertaken until the final climb speed. Depending on the power available from the remaining engines, the aircraft may sustain a shallow climb to ensure adherence to flap placard speeds and improve the altitude over the terrain.

Fourth or *Final segment* begins at the end of the third segment and continues until 1,500 feet above reference zero (to be defined later).

Figure 7-11 shows the general layout of the flight path and the definition of the segments. Also shown are the conditions of the aircraft components and flight parameter values that are prescribed by the FARs, namely, engines, thrust level, gear, flaps, and speed. This picture represents the conditions that must be assumed to calculate the takeoff. For example, the flight path must be calculated such that, among other requirements, the aircraft is flown at V_2 up to at least a 400-foot height, only after which the aircraft can be leveled off and flap retraction initiated.

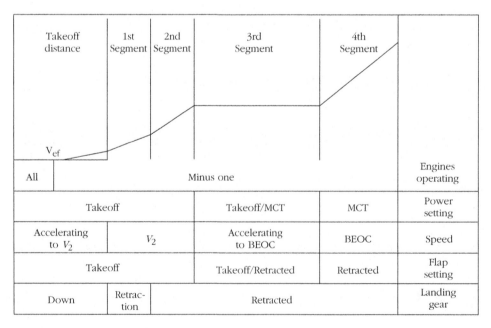

Fig. 7-11 *Schematic of takeoff segments.*

Climb and obstacle limitations

As stated earlier, the TOW of the aircraft might have to be limited to the value that the runway conditions allow. We will now see that the airborne portion of the takeoff can impose limitations on the TOW as well.

Some of these limitations arise as a result of the requirement that the aircraft, if continuing a takeoff after an engine failure, must maintain a minimum gradient of climb. The following table defines the minimum gradient requirements in percent.

Table 7-2

	Segment			
Number of engines	**1**	**2**	**3**	**Final**
2	Positive	2.4	1.2	1.2
3	0.3	2.7	1.5	1.5
4	0.5	3.0	1.7	1.7

The aircraft flight path must meet the above minimum climb gradients or the gradient necessary to satisfy the obstacle clearance requirements, whichever is steeper. It is at this point that we must introduce the concepts of gross and net gradients.

The *gross gradient* is the gradient that the aircraft is capable of flying after an engine failure; Table 7-2 lists gross gradients. The *net gradient* is the gross gradient minus a constant that is specified by the regulations (Table 7-3).

**Table 7-3 Constants used in
defining second-segment net gradient**

Number of engines	**Constant for second segment**
2	0.8%
3	0.9
4	1.0

Thus, for a twin-engine transport, the gross gradient during second-segment climb is 2.4 percent, and the net gradient is 1.6 percent (the result of 2.4–0.8).

The purpose of the net gradient is to provide a safety margin in the calculations of obstacle clearance, which will subsequently be dealt with in this chapter.

Other definitions that will become necessary for the material that follows are (Fig. 7-12 shows some of these concepts pictorially):

- *Reference zero* is the point at which the aircraft reaches 35 feet above the liftoff point.
- *Gross path* is the flight path that complies with the gross-gradient requirement.

- *Gross height* is height above reference zero of any point along the gross path.
- *Net path* is the flight path of the aircraft that complies with the net-gradient requirement.
- *Net height* is the height above reference zero of any point along the net path.

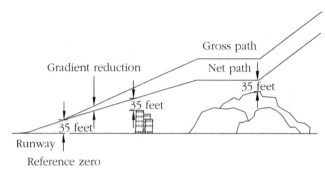

Fig. 7-12 *Schematic of net and gross takeoff paths.*

FAR Part 121.189 requires that the aircraft's net flight path clear all obstacles by 35 feet vertically and 300 feet horizontally; therefore, meeting the basic climb limitations listed in Table 7-2 might not be sufficient to satisfy the obstacle-clearance requirement. Three possible courses of action are available to the operator faced with such a problem:

- Reduce the TOW such that the climb gradient is increased and the net path clears the obstacle by the prescribed 35 feet.
- Modify the takeoff procedure by deviating the takeoff path to fly away from the obstacle. This *special takeoff procedure*, or simply *special procedure*, will be dealt with later in this chapter.
- If the runway conditions and dimensions allow it, increase the takeoff speeds to attain a higher climb gradient. This is called an *overspeed takeoff*, or an *improved-climb takeoff*, and is treated later in the chapter. Note that for obstacle clearance, the improved climb maneuver might be counterproductive since the liftoff point is closer to the obstacle, thereby negating some, if not all, of the benefit gained from the higher climb gradient.

The choice of flap setting has a significant impact on the takeoff process, and its effects require some attention. A higher flap setting shortens the takeoff roll and therefore contributes toward increasing the distance available to climb to a given altitude. *High-lift devices* (HLD) shorten the TOD; but the deployment of higher settings of HLD increases drag and therefore curtails the climb capacity of the aircraft (Fig. 7-13).

If the altitude desired must be reached before reaching point A in Fig. 7-13, then the higher flap setting is desirable. Conversely, if the desired altitude must be

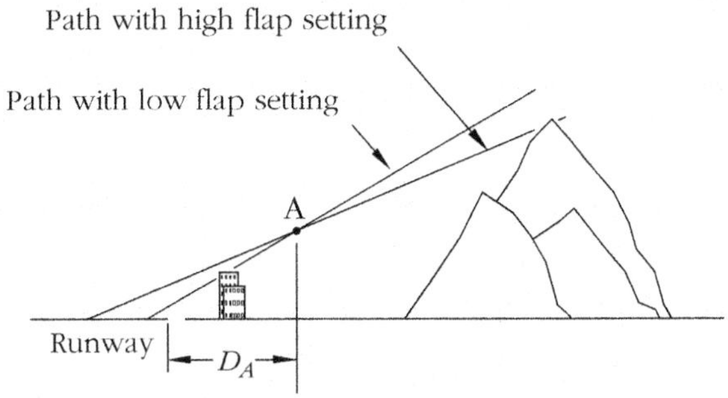

Fig. 7-13 *Effect of HLD on the takeoff path.*

reached after A, then more benefit is derived from the lower flap setting. A similar problem is encountered when designing noise abatement procedures, where noise monitors must be overflown with the lowest possible noise exposure, which might entail flying high over the area, or at low power settings, or both.

Example 7-2

It is known that a twin-engine transport under the prevailing atmospheric conditions and weighing 120,000 lbs. will exhibit a 2.3-percent gross second-segment climb gradient after an engine failure. Knowing (as will be shown in the next chapter) that the climb gradient can be approximated by the expression

$$\gamma = \frac{T - D}{W}$$

What is the maximum weight that can be allowed to meet the climb requirements?

At 120,000 lbs. we know that

$$\gamma_{120} = \frac{T - D}{120,000} = 0.023$$

and the requirement is that

$$\gamma_W = \frac{T - D}{W} = 0.024$$

Dividing one expression by the other, we obtain that

$$\frac{\gamma_W}{\gamma_{120}} = 1.04$$

therefore,

$$W = \frac{120,000}{1.04} = 115,000 \text{ lbs.}$$

What would happen if there is an obstacle 100 feet high at 1 nautical mile from the liftoff point?

Since the aircraft is required to clear all obstacles by 35 feet, the effective height of the obstacle is 135 feet. At 1 nautical mile (6,076 feet) from the end of the runway, this represents a slope of

$$\frac{135}{6076} = 0.0222 = 2.22 \text{ percent}$$

which is only the net gradient. The gross gradient is found by adding 0.8 percent, for a total of 3.02 percent. Now

$$\frac{3.02}{2.3} = 1.33 \Rightarrow W_{obs} = \frac{120{,}000}{1.33} = 91{,}390 \text{ lbs.}$$

These figures are conservative since they assume that the reduction in weight does not have a reduction in drag associated with it (we did not adjust the quantity $T\!-\!D$ during our calculations).

The *overspeed takeoff*, sometimes also known as the *improved-climb takeoff*, is a tool whose purpose is to improve the climb capability of the aircraft. Figure 7-5B shows a condition in which the overspeed takeoff can be used. With the availability of a stopway or a very long runway, the ASD required is greater than the AGD; therefore, for the TOW allowed under the available AGD, there is far more distance to stop the aircraft than necessary. It is conceivable then that the aircraft can be accelerated to a higher V_1 and still have enough stopping distance to bring the aircraft to a safe stop in the event of an aborted takeoff. The same argument with a different picture was illustrated in Fig. 7-10, and naturally so, since what we are exercising here is the availability of multiple V_1 situations, where the V_1 we seek is higher than the V_1 for a BFC, V_{1BAL}.

The benefit of this procedure is that the higher V_1 brings with it the possibility of higher V_R and V_2, and the gradient capability of the aircraft improves with a higher V_2. As will be shown in the chapter on climb, the climb gradient of the aircraft is directly proportional to the difference between thrust and drag:

$$\tan\gamma \approx \gamma = \frac{T-D}{W}$$

Figure 7-14 shows that the difference $T\!-\!D$ is greater for V_{1i} than it is for V_1. Nevertheless, the higher V_1 will be limited by the following considerations:

- The available ASD.
- The increased TOR will bring the aircraft closer to obstacles before becoming airborne, and thereby it might negate the benefit of a higher climb gradient.
- The higher speeds achieved might impinge upon the tire-speed capability or the brake-energy limitations.

Higher speeds, closer to the minimum drag speed and perhaps beyond, would be unlikely to produce any benefit as one of the above limitations would come into play much earlier, not to mention the associated degradation of climb gradient with too high a speed.

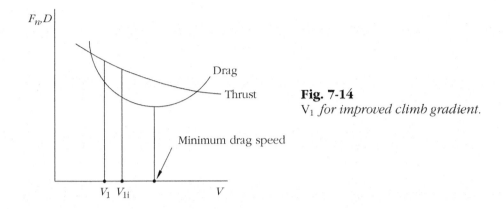

Fig. 7-14
V_1 *for improved climb gradient.*

Obstacle limitations and special procedures

The operator of an aircraft is responsible for ensuring that takeoffs are executed safely and in compliance with the regulations. In addition, the aircraft manufacturer has designed and produced a vehicle capable of operating safely as long as the operator adheres to the procedures and limitations published in the AFM.

The AFM, whether in digital or paper format, is the only certified source of information with which to calculate the aircraft's takeoff performance for any of the conditions under which it was designed to operate. Unfortunately, the AFM, although a source of instructions on how to avoid obstacles vertically, is lacking in similar guidance regarding the avoidance of those obstacles by varying the flight path laterally. But since the safe planning and execution of a takeoff is highly dependent on the identification of the pertinent obstacles, which is the responsibility of the operator, it is important that the operator make use of all available tools in the design of takeoff procedures specifically aimed at avoiding obstacles and thereby maximizing the payload carried aloft.

In the United States, several sources can be used for obstacle identification:
- Sectional charts published by the National Oceanic and Atmospheric Administration (NOAA)
- Obstacle charts and data sheets published by NOAA
- Obstacle data file published by NOAA
- Topographical maps published by the U.S. Geological Survey (USGS)

Unfortunately, these publications, due to their periodic nature, will occasionally exclude obstacles of importance such as construction structures and moving objects such as ships, tall vehicles, and the like. A conscientious appraisal of the local conditions is therefore a mandatory step in the correct identification of all pertinent obstacles.

The essence of a special procedure is the recognition that the aircraft must fly a path designed to avoid obstacles while its performance has been substantially degraded by the loss of power in its most critical engine.

It is very important to understand, and will be underscored here, that the *terminal instrument procedures* (TERPS) normally associated with airports are designed for aircraft operating at normal performance levels. An aircraft that has suffered an engine failure generally will not meet the gradient requirements of a TERPS design, nor is it expected to meet these requirements. TERPS and the special procedures that are the object of this section are completely separate and independent subjects.

As of this writing, the FAA is near the date when a new advisory circular (AC) dealing with obstacle analysis will be published. The essence of this document is to provide a guide to the operator in the elaboration of the special procedures that are necessary to effect obstacle avoidance. The following treatment will draw heavily from this proposed AC. Operators are advised to consult the AC plus local authorities before embarking in the design of a special procedure.

The obstacle accountability area (OAA)

The FARs specify that all obstacles must be avoided either vertically or horizontally:

- The net path must clear an obstacle by 35 feet, or
- The flight path must clear an obstacle by 200 feet horizontally while within the airport boundary or by 300 feet outside it.

The operator cannot be expected to include all obstacles in the vicinity of the airport when preparing a takeoff analysis and design of a special procedure; nor is there a fixed area identified by the regulations within which all obstacles must be considered. There is a need, therefore, to identify an area under and around the intended flight path within which all obstacles will be considered in the design of the special procedure, the *obstacle accountability area* (OAA).

As shown in Fig. 7-15, for a straight-out departure, the edges of the OAA vary linearly from the end of the runway at a ratio of 16-to-1, which means that the resulting splay will have a width of 0.0625L on either side of the intended flight path, where L is the distance along the flight track. Nevertheless, the width

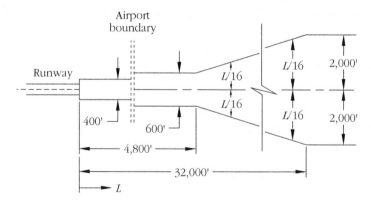

Fig. 7-15 *Splay under a straight out departure.*

of the splay cannot be less than 200 feet on either side of the track within the airport boundary, nor can it be less than 300 feet wide on either side of the track outside the airport boundary. Furthermore, the width of the splay will have a maximum of 2,000 feet on either side of the flight track.

For departures involving turns (Fig. 7-16), any initial straight portion of the track is subject to the same width requirements as a straight-out departure. At the beginning of the turning portion of the track, the width of the OAA is either 300 feet on either side of the track or the width of the previous segment, whichever is greater. Thereafter, the width of the track increases at a rate of $0.125S$ on either side of the track, where S is the distance from the beginning of the first turning segment until reaching a maximum width of 3,000 feet on either side of the track.

There are at least two possible methods of analyzing a departure procedure: the *vertical clearance method* and the *horizontal clearance method*. Each has a particular set of requirements.

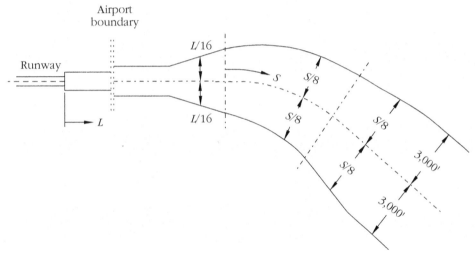

Fig. 7-16 *Splay under a curved departure.*

The vertical clearance method (VCM) of obstacle analysis

The purpose of the VCM is to identify all obstacles that can be overflown while meeting the FAR requirements and still allow the aircraft to exhibit a profitable performance level. At the heart of the method is the OAA, within which all obstacles are to be cleared by the net path with a margin of 35 feet.

For the sake of simplicity, it might be pertinent to use a single intended track that is representative of normal operational procedures, perhaps a standard instrument departure (SID) as an example. This represents few if any problems when the track is straight, the only requirement being that the climb gradient required for the procedure can be attained by the crippled aircraft. For a turning departure, though, it might be necessary to consider several flight tracks because the turning radius of the aircraft will depend on its speed, and the latter, in turn, depends on the takeoff weight. The AC suggests that a single turning track be specified, in which case the bank angle would have to be varied to accommodate the different speeds. The inconvenience of this solution arises in the instructions given to the crew; it is simpler to require the crew to use a single bank angle and design the procedure to protect the aircraft than to request the crew to tailor the bank angle to the takeoff weight.

In the case of multiple tracks, the OAA must be built to encompass all of them by recognizing that the outside tracks define the OAA boundary, as shown in Fig. 7-17.

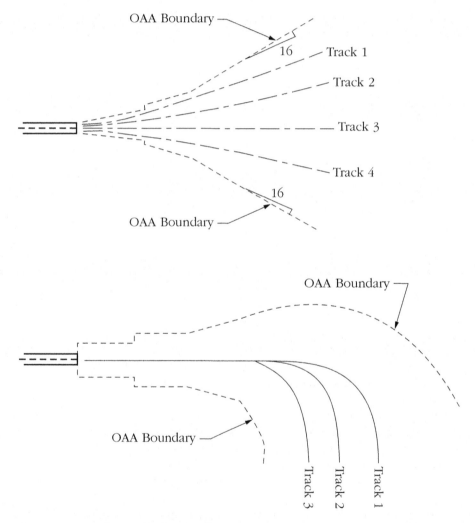

Fig. 7-17 *OAA for multiple takeoff tracks.*

When including an obstacle in the takeoff calculation, the distance to the obstacle will be measured along the centerline of the track, or group of tracks, to a point abeam the obstacle. Naturally, along curved paths, the gradient degradation suffered by the aircraft must be taken into account, and the corresponding penalty applied to the obstacle.

The operator might wish to consider that, under the VCM, there is no requirement to account for the effects of wind, pilot technique, or instrument error on the intended track within the OAA. For example, the 16-to-1 ratio at which the splay expands in a straight-out departure is matched by a 12.5-knot crosswind in a 200-knot (ground speed) departure. Given the possibility of high crosswinds, the operator might want to increase the expansion ratio of the splay, or perform a horizontal clearance method analysis, which is explained in the next section.

The key element in the ability of an aircraft to clear obstacles in its takeoff path is the aircraft speed. As will be shown in the next chapter, for relatively low-speed flight regimes, as is the case during takeoff, the higher the speed, the higher the climb gradient.

Speed increments can be gained by either accelerating to higher speeds during the takeoff roll (overspeed or improved-climb technique) or by accelerating to the desired speed once airborne; the method chosen will depend on the ease with which the aircraft can attain the desired speed, either on the ground or once airborne.

Speed is also necessary to ensure adequate stall margins during turns. In the design of a curved path, the banked turn will carry with it an associated degradation in lift performance (see chapter 6), with two immediate results: the gradient available is degraded and the stall margin of the aircraft is reduced. To alleviate these conditions, the operator might find it necessary to increase the takeoff speeds such that V_2 is higher than normal, and therefore ample enough to accommodate the additional requirements of climb gradient and stall margin.

The speed increases necessary to cover these contingencies are generally provided by the manufacturer in the AFM and should be adhered to.

The horizontal clearance method (HCM) of obstacle analysis

As its name implies, the HCM is intended to avoid obstacles by implementing a flight track around the performance-inhibiting obstacles, not over them.

The AC does not address explicitly the possibility that, in an attempt to fly around tall obstacles, other less-taxing obstacles might have to be overflown anyway, and that these lesser obstacles must be included in the calculation of the takeoff. The tacit question then is to identify these obstacles, and this can be done by remembering that the FARs require that obstacles within 300 feet of the track (outside the airport boundary, 200 feet inside the boundary) be included in the takeoff calculation such that the net path avoids these obstacles by 35 feet vertically.

Because this method does not include the protection of a splay that covers contingencies such as wind, pilot technique, and temperature effect on turns, these must be explicitly addressed by the operator.

It is important to consider the wind that will affect the flight path after liftoff, not only the wind blowing along the airport surface. Figure 7-18 shows a situation in which the crosswind blowing along the runway surface is 10 knots, whereas the wind found at low altitude unmasked by the adjacent hills is 30 knots. The 10-knot crosswind has a minimal effect on the takeoff run, but the 30-knot wind at 100 or 200 feet of altitude might affect considerably the ground track of the aircraft in the event that an engine fails and a turn is initiated to avoid the distant hill along the runway heading. Hopefully, in situations such as this, the airport authorities have wind-measuring equipment located such that not only the surface winds but the winds at low altitudes are available for flight crew briefings.

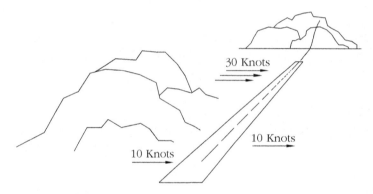

Fig. 7-18 *Winds on the surface and at a selected altitude above the runway.*

Chapter 6 detailed the effects of wind on the turning path of the aircraft. The application of the principles illustrated there facilitates the design of special procedures when crosswinds must be considered.

The effect of pilot technique is important in the design of a procedure that inherently includes turns to avoid obstacles. It must be kept in mind that the pilot executing this procedure will be under considerable stress—the aircraft has experienced an emergency that might remain unresolved while the pilot must continue flying a crippled aircraft, possibly in less than benign weather. Under these circumstances, the ability of a pilot to hold a bank angle accurately is impaired, and, in turn, the resulting turn radius and gradient degradation are affected. The results of these uncertainties are best assessed in a simulator, perhaps with the help of a human factors expert, where different scenarios can be attempted, perhaps resulting in a redesign of the special procedure.

Figure 7-19 shows a plot-of-turn radius against temperature and KCAS, the latter being the speed used in the definition of a special procedure. In designing the procedure, it is important to consider the seasonal variation of temperature and its effect on TAS and therefore on the turn radius. For example, from the plot, the difference in radius of turn at 170 KCAS is some 2,000 feet between a cold and a warm atmosphere; this could be the difference between including an additional obstacle or avoiding it.

Course guidance

For VCM and HCM analysis, course guidance might be available to enhance the certainty with which the location of the actual flight track is known relative to the intended flight track. With ground-based course guidance, for example, the OAA and the number of obstacles that need to be included in the takeoff calculation can be reduced considerably.

The AC specifies the allowances associated with each type of course guidance available to the operator; we will only mention, as an example, the allowances pertinent to localizer approach equipment. For a localizer, the AC specifies that the splay to be utilized spreads at a 1.25° angle, with a minimum half width of 300 feet covering the distance from the LOC out to 2.25 nautical miles away (Fig. 7-20).

Other forms of ground-based guidance are DME, ADF, and VOR. Similarly, some credit can be taken for guidance provided by onboard equipment such as flight management computers, as well as space-based equipment, such as the global positioning system.

Visual guidance is another valid method of navigating, too; but the AC specifies that an HCM analysis must be performed to use visual guidance.

Other takeoff limitations

Previous sections considered several possible limitations to takeoff: field length, climb capability, and obstacle clearance. Additionally, during the takeoff roll, the aircraft must not exceed the speed beyond which the tires will suffer structural failure, nor the speed beyond which the brakes cannot absorb the kinetic energy of the aircraft. A further limitation, one that is becoming more common, is noise abatement.

Additionally, other sources of reductions in the takeoff performance of the aircraft stem from the equipment that the aircraft is assumed to have to execute a normal takeoff. Documents such as the minimum equipment list, configuration deviation list, or the like, define the penalties associated with anomalies in the aircraft. For example, a missing flap seal might call for a mandatory increase in takeoff speeds, a reduction of a few hundred pounds in takeoff weight, or both to account for the degraded aerodynamics. Likewise, a deactivated antiskid brake system will probably result in a substantial increase in the required ASD,

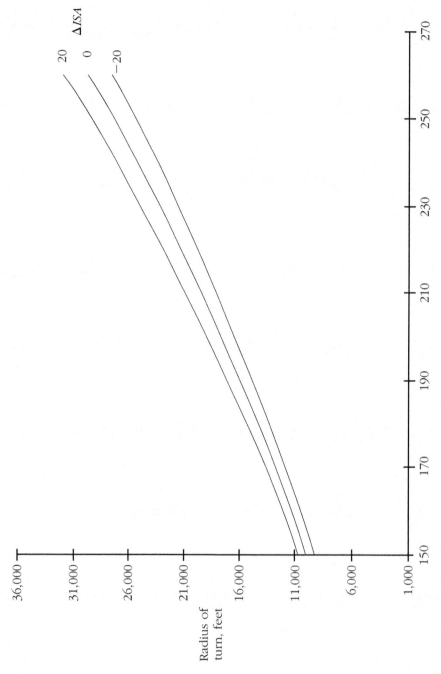

Fig. 7-19 *Effect of temperature on turn radius.*

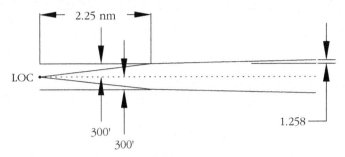

Fig. 7-20 *Splay for a localizer departure.*

with the attendant reduction in takeoff weight. These penalties, of course, are specified in the AFM.

All of these considerations limit the takeoff weight of the aircraft, and the most restrictive limitation actually defines the allowable takeoff weight. The process of takeoff analysis consists of:

- Analyzing in detail the environment in which the takeoff is to be made.
- Determining the allowable takeoff weight (utilizing techniques such as the ones presented here).
- Specifying the takeoff flight path to be followed.

Takeoff analysis is an iterative process, one that visits each and every source of takeoff-weight limitations and adjusts the takeoff parameters in a continuous search for the maximum takeoff weight capability; therefore, a thorough and efficient takeoff analysis capability is of paramount economic importance to the fleet operator.

Takeoff equations

Newton's Second Law states that the acceleration of a body is directly proportional to the forces acting on it and inversely proportional to its mass:

$$F = ma \tag{7-1}$$

Furthermore, from basic dynamics we know that

$$a = \frac{dV}{dt}; \qquad V = \frac{ds}{dt} \tag{7-2}$$

To apply these concepts, consider the diagram shown in Fig. 7-21, where the lift force L is the total lift force produced by all the aerodynamic surfaces. Note that this lift is produced while the aircraft is rolling and is subject to *ground effect*; it is therefore incorrect to calculate this lift force from the C_L given in the manuals for the aircraft in flight.

Similarly, the drag force D is the aerodynamic drag generated by the airframe. Like lift, drag is subject to ground effect and cannot be calculated as is normally done for an aircraft in flight. (An alternative method of calculation for drag and lift will be treated later.)

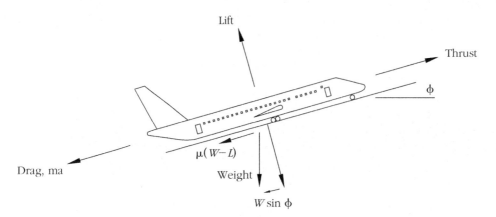

Fig. 7-21 *Forces acting on the aircraft during the takeoff roll.*

The friction force *F* arises from the resistance to motion imparted by the deflection of the tires and bearing friction within the axles. In this context of rolling friction, μ is not the coefficient of friction between the tires and the pavement; in other words, it is not the same μ that would be used in braking calculations where slippage between the tire and the pavement is important.

The weight *W* of the aircraft acts perpendicular to the Earth, not to the local surface, which might be inclined at an angle ϕ; ϕ is the angle between the runway surface and the local horizontal. In the calculation of takeoff, the angle ϕ is defined by the difference in elevation between opposite ends of the runway to be used. In Fig. 7-22, ϕ would be calculated as if the runway were the line joining A to C; in reality the aircraft might have a takeoff roll from A to B, in which case the slope is of opposite sign to that prescribed by the regulations. Although inaccurate, this convention has been widely adopted throughout the industry.

Slopes are considered negative if the aircraft rolls uphill and positive if the aircraft rolls downhill for takeoff. Furthermore, in dealing with the equations of motion, we treat ϕ as an angle (although often approximated by its sine or tangent functions), whereas the data for a runway is usually given in percent of slope, the tangent of the angle.

Applying Newton's Second Law (which states, roughly, that the sum of the forces equals the product of the mass and the resulting acceleration):

$$\sum F = ma = \frac{W}{g}a$$

and considering all the forces described above,

$$T - D - \mu(L - W) - W\sin\phi = \frac{W}{g}a \qquad (7\text{-}3)$$

where the $\sin\phi$ can be replaced by ϕ expressed in radians, thereby simplifying the equation.

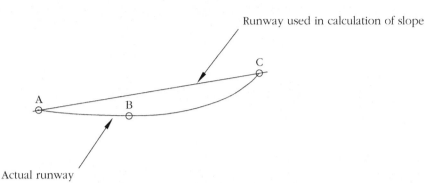

Runway used in calculation of slope

Actual runway

Fig. 7-22 *Definition of runway slope.*

Solving for a and substituting Equations 3-6 and 3-7 results in

$$a = \frac{g}{W}[(T - \mu W) - (C_D - \mu C_L)Sq - W\phi] \tag{7-4}$$

During flight tests, the manufacturer establishes the value of the quantity $(C_D - \mu C_L)$, which, for the most part, stays constant during the takeoff roll. Substituting into Equation 7-2 and integrating to solve for V:

$$V = \int_{t_1}^{t_2} a.dt + V_w = \frac{g}{W}\int_{t_1}^{t_2}[(T - \mu W) - (C_D - \mu C_L)Sq - W\phi] \, dt + V_w \tag{7-5}$$

where V_w is the wind velocity. Integrating once again to find the distance covered,

$$s = \int_{t_1}^{t_2} V.dt \tag{7-6}$$

where s_0 is the initial position of the aircraft at the brake-release end of the runway.

The solution of the integral in Equation 7-6 requires a detailed knowledge of the behavior of T, W, and q as the velocity of the aircraft increases. Only rarely will such an integral be solved in closed form. More likely the calculation will be performed in steps with the assistance of a computer.

Throughout this development we have been using an expression for acceleration that depends on the drag coefficient C_D. It is important to underscore the effect of high-lift devices (flaps, slats, etc.) on C_D and, through it, on acceleration. Obviously, when more surfaces are exposed to the relative wind, more drag will be produced.

High-lift devices (HLD), as the name implies, increase the lift capability of the aircraft and will thus decrease V_{LOF}; consequently, the distance required to lift off is reduced. High-lift devices on the one hand increase the resistance to motion and on the other hand enhance the lift capability of the aircraft. The benefit accrued from increased lift is usually more than the degradation in performance stemming from the additional drag; the net result is a shortened TOD with the higher flap settings. Also, in reducing V_{LOF} and V_1 by using higher flap settings, care must be exercised not to violate the constraint that $V_1 \geq V_{mcg}$. It is also important to remember that the higher flap settings inhibit the climb capability of the aircraft, which might be an important consideration in clearing obstacles, or in meeting a noise restriction.

8

Climb

The climb formulation developed in this chapter will be general in nature and applicable to takeoff calculations and enroute climb-performance problems. When considering subsequent enroute climb performance, simplifying assumptions will be made to render the problem more tractable.

The emphasis in this chapter is necessarily analytical because the subject is a portion of the flight in which both the energy of the aircraft and the atmospheric conditions are changing relatively rapidly. For those readers whose needs are not as much quantitative as qualitative, the subsections on speed effects, altitude effects, angle of climb, and economy climb will be of particular interest.

Climb equations

The equations developed below were derived by considering the forces acting on the climbing airplane, as shown in Fig. 8-1, where the summation of forces will be done along the direction of the relative wind, or the aircraft velocity vector. Furthermore, since these equations are generally solved numerically, the rate of change of weight (fuel burn off) is not considered, under the assumption that the aircraft weight will be updated at each step along the integration process.

Summing the forces along the x-direction yields

$$\sum F_x = ma \Rightarrow \tag{8-1}$$

$$ma = T\cos\alpha - D - W\sin\gamma \Rightarrow$$

$$\frac{\dot{V}W}{g} = T\cos\alpha - D - W\sin\gamma \Rightarrow$$

$$\dot{V} = g\left[\frac{T\cos\alpha - D}{W} - \sin\gamma\right]$$

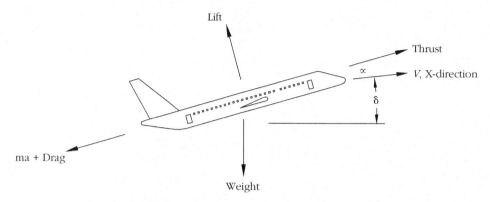

Fig. 8-1 *Forces acting on the aircraft during the climb.*

The summation of forces along the z-axis, perpendicular to the x-axis, produces the following equations:

$$\sum F_z = ma_z \Rightarrow$$
$$ma_z = mVq = L - W\cos\gamma + T\sin\alpha$$

where q is the pitch rate, or the rate of change of θ (θ is the sum of γ and α). Incorporating these concepts results in the following equation:

$$\frac{W}{gV}\frac{d}{dt}(\alpha + \gamma) = L - W\cos\gamma + T\sin\alpha$$

which is useful in dealing with climb conditions such as those found shortly after liftoff. For the less severe case of enroute climb, the pitch rate can be assumed to be zero:

$$0 = \frac{L + T\sin\alpha}{mg} - \cos\gamma \tag{8-2}$$

Rate of climb

The rate of climb (ROC) is the vertical projection of the aircraft velocity; accordingly, ROC will be positive during a climb and negative during a descent. ROC can be expressed as the product of the forward velocity and the sine of the climb angle, γ; or as the rate of change of altitude:

$$ROC = V_T\sin\gamma = \frac{dh}{dt} \tag{8-3}$$

Assuming that α is small such that $\cos\alpha$ is close enough to 1, Equation 8-1 can be rewritten:

$$\frac{dV_T}{dt} = g\left(\frac{T - D}{W} - \sin\gamma\right)$$

This small angle approximation is consistent with flight during the enroute portion of the climb to cruise altitude, or even during the takeoff, in the case of an engine failure.

We can also write

$$\frac{dV}{dt} = \frac{dV}{dh}\frac{dh}{dt} = \frac{dV}{dh}ROC$$

and therefore, from the previous equation:

$$V_T\frac{dV_T}{dh}ROC = g\left[V_T\frac{T-D}{W} - ROC\right] \tag{8-4A}$$

or, rearranging:

$$ROC = \frac{V_T\dfrac{T-D}{W}}{1 + \dfrac{V_T}{g}\dfrac{dV_T}{dh}} \tag{8-4B}$$

The projection of V_T on the ground is

$$V_T\cos\gamma$$

which, when combined with the wind velocity, V_W, will yield the ground speed, V_g:

$$V_g = V_T\cos\gamma + V_W \tag{8-5}$$

If γ is sufficiently small, then we can assume that $\cos\gamma$ is approximately equal to 1, and substitute into Equation 8-4 to obtain

$$ROC = \frac{V_T\dfrac{T-D}{W}}{\left[1 + \dfrac{V_T}{g}\left(\dfrac{dV_T}{dh} + \dfrac{dV_W}{dh}\right)\right]} \tag{8-6}$$

This expression for ROC reveals the influence exerted by TAS, drag, thrust, weight, and the change in wind with altitude. The actual expression used in calculations is Equation 8-4B; Equation 8-6 merely underscores that it is the rate of change of wind with altitude, not just the wind itself, that affects the rate of climb.

The term

$$\frac{V_T}{g}\left(\frac{dV_T}{dh} + \frac{dV_W}{dh}\right) \tag{8-6A}$$

in Equation 8-4B appears often in calculations involving climb and descent trajectories, it is therefore given a special name: *acceleration factor.* Its significance arises from the possibility that the aircraft's propulsive power can be used to effect a change in altitude, a change in speed, or a combination of both. The acceleration factor is a device that helps the bookkeeping in calculations involving an accelerated climb or descent. Note, therefore, that in an unaccelerated climb, ROC reduces to

$$ROC = V_T \left[\frac{T - D}{W} \right] \qquad (8\text{-}7)$$

Again, Equation 8-6 shows that ROC is not affected by wind, but by the rate of change of wind with altitude. This might seem contrary to experience, when a strong headwind would seem to increase the aircraft's capability to climb. Our perception is one of angle, not rate of climb; and angle of climb is indeed affected by wind, as described in the subsection on climb angle.

Another version of Equation 8-4A is

$$(T - D)V_T = W\frac{V_T}{g} \frac{dV_T}{dh} ROC + W \cdot ROC$$

where the term in the left-hand side of the equation is called the *excess power*, the first term in the right-hand side is the *kinetic energy rate*, and the last term is the *potential energy rate*:

$$Excess\ Power = \frac{d(KE)}{dt} + \frac{d(PE)}{dt}$$

If the above expression is divided through by the weight, the equation becomes one of specific power in terms of specific energy. Either form is used mostly in the study of high-performance aircraft such as fighters.

Traditionally, climb schedules have been designed to maximize ROC; and, coincidentally, the maximum ROC schedule closely follows a line made up of a constant CAS segment and a constant Mach segment as shown in Fig. 8-2. Constant CAS/Mach schedules have been widely adopted by the industry for climb and descent velocity profiles; they offer simplicity of implementation and operation for both the air carriers and air traffic control.

Unfortunately, these simple schedules offer some complications for the analyst. A constant CAS velocity profile, for example, has associated with it a continuously changing TAS; in other words, there is an acceleration associated with a constant CAS climb. Similarly, a constant Mach climb will entail a decreasing TAS. There is a need then to investigate the nature of the acceleration factor in constant CAS/M climb profiles.

Acceleration factor corrections

We will derive an expression for acceleration factor in a constant-Mach climb. Similar derivations can be worked for constant CAS and constant EAS schedules.

We have defined acceleration factor as

$$\frac{V_T}{g} \left(\frac{dV_T}{dh} + \frac{dV_W}{dh} \right) \qquad (8\text{-}8)$$

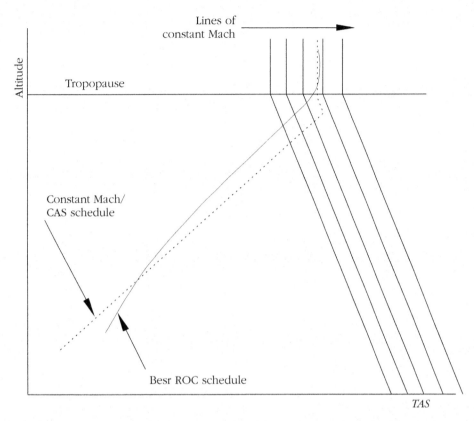

Fig. 8-2 *The normal climb schedule approximates a best* ROC *climb schedule.*

Using Equations 2-6 and 3-5, we can derive the following expression for the true air speed, V_T:

$$V_T = Ma = M\sqrt{\gamma gRT} \tag{8-9}$$

where "a" is the local speed of sound, and "T" is the ambient absolute temperature. From the material in chapter 1, we can derive an expression for ambient temperature in terms of the temperature lapse rate λ

$$T = T_0 - \lambda h_p$$

Using this form in Equation 8-9,

$$V_T = Ma = M\sqrt{\gamma gR}\sqrt{T_0 - \lambda h_p} \tag{8-10}$$

therefore,

$$\frac{dV_T}{dh_p} = M\sqrt{\gamma gR}\frac{(-\lambda)}{2\sqrt{T_0 - \lambda h_p}} = -\lambda M\frac{\sqrt{\gamma gR}}{2\sqrt{T_0 - \lambda h_p}}$$

Multiplying through by V_T,

$$V_T \frac{dV_T}{db_p} = (M\sqrt{\gamma gR}\sqrt{T_0 - \lambda b_p})\left(\frac{-\lambda M\sqrt{\gamma gR}}{2\sqrt{T_0 - \lambda b_p}}\right)$$

which simplifies to

$$V_T \frac{dV_T}{db_p} = -\frac{M^2 \gamma gR\lambda}{2}$$

or,

$$\frac{V_T}{g}\frac{dV_T}{db_p} = \left(\frac{\gamma M^2}{2}\right)(-\lambda R) \qquad (8\text{-}11)$$

Table 8-1 contains a summary of the acceleration factor forms that apply for different speed schedules.

Table 8-1 Climb schedule

	Constant M	**Constant EAS**	**Constant CAS**
Above tropopause	0	$0.7M^2$	K
Below tropopause	$-0.133M^2$	$0.567M^2$	$-0.133M^2 + K$
	$K = (1 + 0.2M^2) - (1 + 0.2M^2)^{-2.5}$		

In a ROC calculation at constant M, below the tropopause, and with no wind, for example, we would substitute the appropriate term ($-0.133M^2$) from Table 8-1 into Equation 8-6:

$$ROC = \frac{V_T \dfrac{T - D}{W}}{[1 - .133M^2]}$$

Example 8-1

Suppose that we choose to study the unaccelerated ROC for an aircraft at a value of 2,000 fpm. Assuming there is no wind gradient, how would this value of ROC change for a CAS = 320 KCAS climb at 20,000, 25,000, and 37,000 feet? What about a Mach = 0.720 climb at the same altitudes?

For the solution, we resort to Equation 8-6, knowing that the acceleration factor term (AF) shown in Equation 8-6A is to be obtained by one of the expressions in Table 8-1:

$$ROC = \frac{V_T \dfrac{T - D}{W}}{1 + AF} \qquad (8\text{-}4B)$$

(1.) Consider first the constant M schedules:
For a constant M climb below the Tropopause (36,089')
$AF = -0.133M^2$; therefore,
$AF = -0.133(.720)^2 = -0.0689$

We already know that the unaccelerated ROC is 2,000 fpm,

$$V_T \frac{T-D}{W} = 2,000 \; fpm$$

and there is no need to calculate thrust or drag.

We can now write,

$$ROC = \frac{2000}{1 - 0.0689} = 2148 \; fpm$$

For the case of 37,000', $AF = 0$ indicating that the ROC remains unchanged at 2,000 fpm.

(2.) For the constant CAS schedules:

First, since Table 8-1 gives the AF in terms of M, we need to convert the CAS values to M using Equation 2-5A:

CAS = 320 @ 20,000' $\Rightarrow$ $M = 0.693$
CAS = 320 @ 25,000' $\Rightarrow$ $M = 0.761$
CAS = 320 @ 37,000' $\Rightarrow$ $M = 0.962$

Even though the last value of M (0.962) is probably too high for any existing transport aircraft, we will continue to use it to illustrate the principles involved.

We now calculate the parameter K, as defined in Table 8-1, for each of the above M values:

$M = 0.693 \Rightarrow K = 0.3009$
$M = 0.761 \Rightarrow K = 0.3555$
$M = 0.962 \Rightarrow K = 0.5310$

The values of AF can be obtained:
CAS = 320 @ 20 kft $\Rightarrow$ $M = 0.693 \Rightarrow AF = K - 0.133M^2 = 0.2370$
CAS = 320 @ 25 kft $\Rightarrow$ $M = 0.761 \Rightarrow AF = K - 0.133M^2 = 0.2785$
CAS = 320 @ 37 kft $\Rightarrow$ $M = 0.962 \Rightarrow AF = K - 0.0 \;\;\; = 0.5310$

Finally, the ROC can be calculated:
CAS = 320 @ 20 kft $\Rightarrow$ ROC = 2,000/(1 + 0.2370) = 1,617 fpm
CAS = 320 @ 25 kft $\Rightarrow$ ROC = 2,000/(1 + 0.2785) = 1,564 fpm
CAS = 320 @ 37 kft $\Rightarrow$ ROC = 2,000/(1 + 0.5310) = 1,306 fpm

Additionally, let's calculate the ROC for 22,000 feet: immediately we notice that, in determining the value of M that corresponds to 320 KCAS at 22,000 feet, the resulting value is precisely 0.72! Indeed, for every altitude, there is a combination of CAS/M, both of which correspond to the same TAS. Conversely, for every combination CAS/M, there is a unique altitude where both speed measures yield the same TAS; this altitude is known as *cross-over altitude*. This is also evident from Equation 2-5, where we can elect to vary either CAS or M to obtain δ, which represents altitude.

Solving first as a constant M problem, we derive the same result as before, namely ROC = 2,148 fpm.

As a constant CAS formulation, first we find $K = 0.3222$.
Then $AF = .3222 - 0.133(0.72)^2 = 0.2533$
And finally ROC $= 2,000/(1 + 0.2533) = 1,595$ fpm.

Summarizing our results:

	Const. M	Const. CAS
37,000'	2,000	1,420
25,000'	2,148	1,563
22,000'	2,148	1,595
20,000'	2,148	1,617

Important points to observe:

- For a constant M climb below the tropopause, the ROC will remain constant since, at a constant M speed schedule, the aircraft is continuously decelerating to match the continuously decreasing speed of sound with increasing altitude.
- In a constant CAS climb, the TAS continuously increases with altitude, in essence requiring that the aircraft accelerate during the climb; consequently, the ROC will decrease as altitude increases.
- At the cross-over altitude, even though the TAS is the same for both speed schedules, the ROC will be different, reflecting the different acceleration states of the aircraft.
- We have used here an example where the unaccelerated ROC was fixed at 2,000 fpm. This serves to illustrate some of the concepts associated with acceleration-factor corrections but might not be representative of the actual performance of an aircraft, particularly at the higher altitudes where the thrust and drag dependence on altitude yield a much curtailed climb capability.

Speed effects

If we divide Equation 8-3 by the velocity V,

$$\frac{ROC}{V} = \frac{V\sin\gamma}{V} = \sin\gamma$$

We have as a result an expression that allows us to conclude that the larger the ratio of ROC to airspeed, the larger the climb angle γ. Figure 8-3 presents a pictorial representation of the same idea. The curve shown is a typical plot of ROC versus speed, reaching a maximum where the quantity

$$\frac{V(T-D)}{W}$$

reaches a maximum (see Equation 8-6).

Note also that the slope of the straight line is precisely ROC/V and that the maximum slope is found at a speed somewhat lower than that for maximum ROC. At this point, we find the speed for maximum climb angle. Reference to

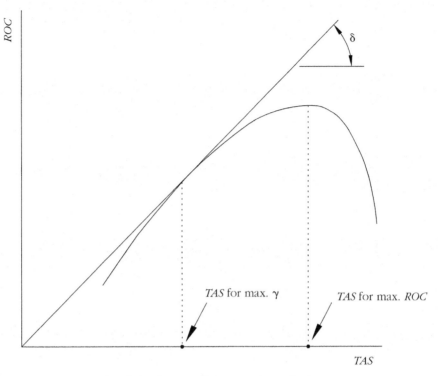

Fig. 8-3 *Best angle of climb,* γ, *and the speed at which it occurs.*

Fig. 8-4 reveals the behavior of this curve as the climb performance of the aircraft degrades (due to higher altitudes, or higher weights).

Note that as long as the maximum ROC is greater than zero, the velocity for the maximum gradient is lower than the velocity for maximum ROC. Eventually the peak ROC becomes zero at a speed that coincides with the speed at which the maximum angle also reaches a value of zero; thereafter, the relationship between maximum ROC speed and maximum angle speed reverses, whereby the velocity for maximum ROC (minimum rate of descent) will be higher than the speed for maximum angle of climb (minimum angle of descent).

Altitude effects

As altitude increases, thrust available and thrust required decrease, but not at the same rate; the thrust available will decrease at a faster rate, reaching a point where the difference between these two quantities becomes zero, and along with it ROC (Fig. 8-5). At this point, the aircraft has reached its absolute ceiling.

The terms absolute ceiling and service ceiling are often used in the context of describing an aircraft's ability to climb. Figure 8-6 presents a plot of thrust and drag versus velocity for several altitudes and the associated ROC values. As altitude increases, the value of the quantity $(T - D)/W$ decreases, and with it the

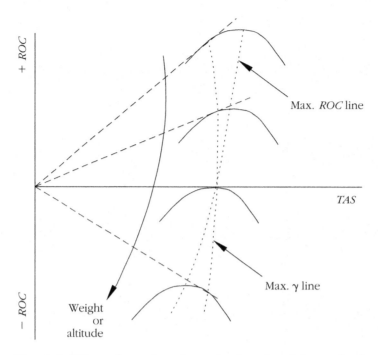

Fig. 8-4 *Effect of speed and weight (or altitude) on climb performance.*

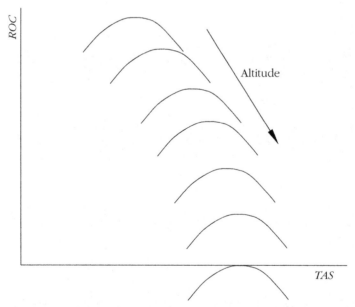

Fig. 8-5 *Effect of speed and altitude on climb performance.*

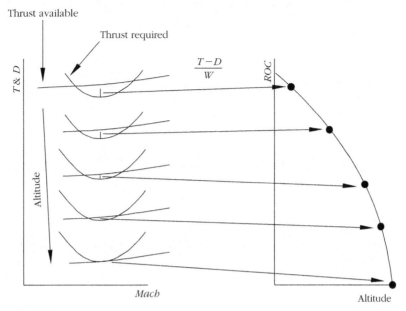

Fig. 8-6 *Thrust, drag, and* ROC.

ROC. The altitude at which the ROC becomes zero is called the *absolute ceiling.* The *service ceiling* is the altitude at which the ROC has some residual value, usually 100 to 300 fpm.

Angle of climb and climb gradient

The climb gradient is formally defined as the tangent of the climb angle,

$$gradient = \tan\gamma$$

Practically, the climb angles encountered in commercial operations during the enroute climb portion of the flight are small enough to allow some simplifications,

$$\tan\gamma \approx \sin\gamma \approx \gamma$$

for γ expressed in radians. Consequently, we often see in the literature that γ as well as $\sin\gamma$ are also called the climb gradient.

If Equation 8-6 is divided by V_T, the resulting expression is the ratio of the vertical component of Fig. 8-7 to the hypotenuse of the velocity triangle, or the sine of the angle γ

$$\frac{ROC}{V_T} = \frac{\dfrac{T-D}{W}}{\left[1 + \dfrac{V_T}{g}\left(\dfrac{dV_T}{dh} + \dfrac{dV_W}{dh}\right)\right]} = \sin\gamma \qquad (8\text{-}12)$$

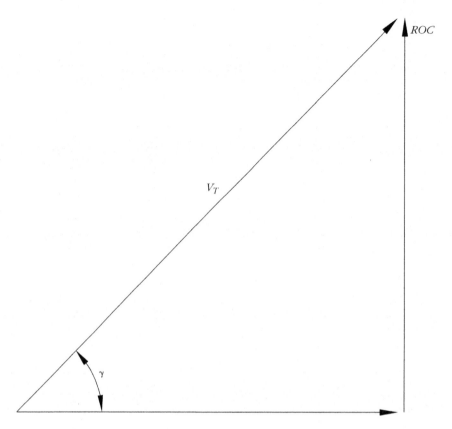

Fig. 8-7 *Geometry of true airspeed (TAS) and climb speed (ROC) vectors.*

The angle of climb depends also on the rate of change of V_T with altitude. Unfortunately, these results are deceptive.

The ground-relative climb angle γ_g is measured with respect to an inertial system of reference axes, as seen by an observer on the ground. It is this angle that will tell us if obstacles will be cleared or if a given altitude will be reached before a certain distance. γ_g is affected by wind directly.

From an aircraft in the air, it is impossible to tell the speed of the wind by purely aerodynamic measurements (no accelerometers, no electromagnetic emissions). Consequently, the velocities associated with the aircraft in flight can be shown as in Fig. 8-8, even if the aircraft is flying into a constant headwind (Fig. 8-9). Note that the velocity V of Fig. 8-9 is not the same as the velocity V_T of Fig. 8-8. V is measured with respect to the ground, and V_T is measured with respect to the air mass. Figure 8-10 shows the velocity triangle that will be drawn by a ground observer. To this observer, the wind velocity is very much in evidence, as is the ROC.

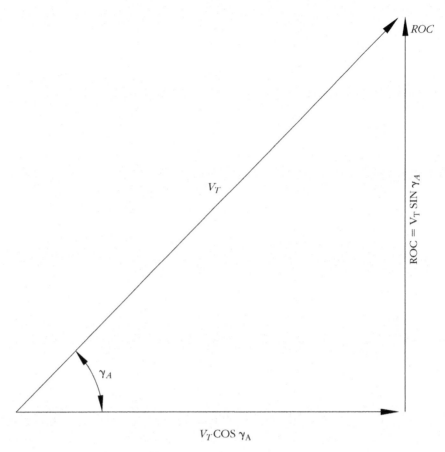

Fig. 8-8 *Air-relative velocity triangle.*

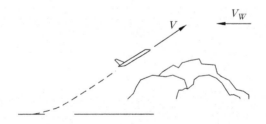

Fig. 8-9 *Climb into a headwind.*

ROC is unchanged when viewed either from the ground or from the air; it is not affected by wind, as was also borne out by Equation 8-6. It is to be expected then that the lengths of the vectors representing ROC in Figs. 8-8 and 8-10 are the same.

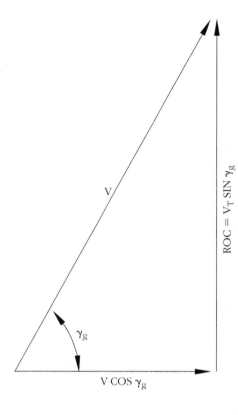

$$\text{ROC} = V_T \sin \gamma_g$$

$$V$$

$$\gamma_g$$

$$V \cos \gamma_g$$

Fig. 8-10
Ground-relative velocity triangle.

The same is not true of what the ground observer perceives as the ground speed. In the presence of strong headwinds, it seems that aircraft travel slowly, when in reality their speed with respect to the air mass is quite sufficient. Figure 8-8 shows that the horizontal component of velocity is

$$V_T \cos\gamma_a$$

This velocity must necessarily be the same as what the ground observer calls ground speed

$$V\cos\gamma_g - V_W$$

Equating these two speeds as prescribed

$$V_T\cos\gamma_a = V\cos\gamma_g - V_W \qquad (8\text{-}13)$$

To investigate the relative magnitudes of the climb angles γ_g and γ_a, we will construct their tangents by referencing Figs. 8-8 and 8-9

$$\tan\gamma_a = \frac{ROC}{V_T\cos\gamma_a}$$

$$\tan\gamma_g = \frac{ROC}{V_T\cos\gamma_g}$$

but, from Equation 8-13, we can also write this last expression as

$$\tan\gamma_g = \frac{ROC}{V_T\cos\gamma_a + V_W}$$

Both tangents together are then

$$\tan\gamma_g = \frac{ROC}{V_T\cos\gamma_a + V_W} \qquad \tan\gamma_a = \frac{ROC}{V_T\cos\gamma_a}$$

Considering that the value of V_W is negative for a headwind, we can conclude that in the presence of headwind, the tangent of γ_g is larger than the tangent of γ_a, indicating that γ_g is larger than γ_a, which matches our perception of reality.

Acceleration gradient

From Equations 8-3 and 8-6, we can derive an expression for the climb gradient; dividing Equation 8-6 by V_T,

$$\gamma \approx \sin\gamma = \frac{\dfrac{T-D}{W}}{\left[1 + \dfrac{V_T}{g}\left(\dfrac{dV_T}{dh} + \dfrac{dV_W}{dh}\right)\right]}$$

Also, from summing forces along the longitudinal axis in Fig. 8-11 we have

$$T - D = \frac{W}{g}a \Rightarrow \frac{T-D}{W} = \frac{a}{g}$$

We see here that the quantity

$$\frac{T-D}{W}$$

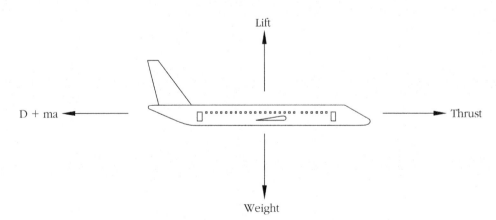

Lift

D + ma

Thrust

Weight

Fig. 8-11 *Forces acting on the aircraft during accelerated flight.*

can be considered a potential that can be transformed into either acceleration or climb gradient.

We can therefore write (somewhat arbitrarily)

$$\frac{T-D}{W} = \sin\gamma\left[1 + V_T\left(\frac{dV_T}{db} + \frac{dV_W}{db}\right)\right] + \frac{a}{g} \qquad (8\text{-}14)$$

or,

TOTAL GRADIENT = DESIRED GRADIENT + ACCELERATION RATIO

where, by acceleration ratio we mean acceleration expressed in Gs.

The total gradient potential, $(T-D)/W$, then can be expended to achieve some value of desired gradient, or some value of acceleration, or a combination of both.

Two points must be emphasized about Equation 8-14:
- The equation applies for any aircraft, so that if we plot desired gradient against total gradient for several values of acceleration (Fig. 8-12), the graph can be used for any aircraft.
- It is an algebraic balance. The equation does not say that climb gradient and acceleration can be traded instantaneously. It does say that, under the prevailing dynamic state of the aircraft, acceleration and climb gradient are available in the relative proportions established by the equation. The transfer of acceleration into gradient or vice versa must still be achieved by maneuvering the aircraft over a finite period of time.

In the absence of wind, the gradient equivalence equation becomes

$$\gamma_{total} = \frac{a}{g} + \gamma_{desired} \Rightarrow \gamma_{desired} = \gamma_{total} - \frac{a}{g}$$

Example 8-2

Suppose an aircraft exhibits a total gradient of 3 percent, and is currently climbing at a 2-percent gradient. What is the maximum acceleration available to the aircraft while maintaining the current climb gradient?

$$\frac{a}{g} = \gamma_{total} - \gamma_{desired} \qquad \frac{a}{g} = 0.03 - 0.02 = 0.01$$

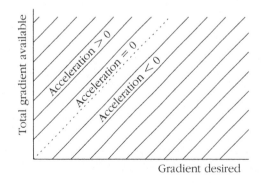

Fig. 8-12
Trade off between acceleration and climb gradient.

Therefore, 0.01 g of acceleration is possible, or 0.322 ft/sec², which equals about 0.19 kts/sec. To gain some perspective, note that at this acceleration it will take the aircraft some 52 seconds to gain 10 knots.

Climb performance with one engine inoperative

Two variations to the engine-inoperative climb calculation are:
- Flight at fixed power
- Flight at thrust required power

In the first case, the power is set to a selected power setting, usually MCT rating, the maximum continuous power allowed under the prevailing atmospheric and dynamic conditions—a power level that will allow the aircraft to either accelerate or to maintain the maximum ROC possible.

In the second case, the power is set to the level required to maintain a desired ROC or speed—under the assumption that the power available is at least as great as the power required.

With reference to Fig. 8-13, a summation of forces along the longitudinal axis of the aircraft will result in the following expression:

$$nT_n = W\sin\gamma + D \qquad (8\text{-}15)$$

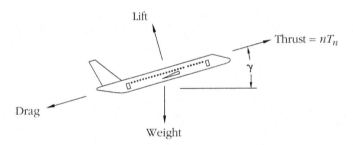

Fig. 8-13 *Forces acting on the aircraft in an unaccelerated climb.*

where, in essence, we have assumed that the angle of attack, α, is negligibly small, an assumption that would not be valid in a takeoff condition but acceptable once the aircraft has completed the second segment and proceeds to accelerate and/or clear the obstacles in its path. Also, in Equation 8-15, the drag D is the total drag of the airframe; it is made up of the base drag with which we have worked extensively so far, plus the yaw drag and windmilling drag (or locked rotor drag, if the engine is not spinning freely). The variable "n" is the number of engines still operating, each producing a thrust "T."

Yaw drag results from both the deflection of the rudder to counter the asymmetrical thrust and the additional drag incurred by virtue of the asymmetrical attitude of the airframe with respect to the velocity vector, an attitude measured by the angle of sideslip, β.

Windmilling drag is the drag associated with an engine that is not only producing no thrust, but whose rotors are forced to rotate by the onrushing wind.

Equation 8-15 can now be rewritten

$$nT = W\sin\gamma + D + D_{WM} + D_{YAW}$$

or, in terms of aerodynamic coefficients:

$$nT = W\sin\gamma + C_D qS + C_{D_{WM}} qS + C_{D_{YAW}} qs$$

where "q" is the dynamic pressure (see Equation 2-1). The yaw drag coefficient, C_{DYAW}, is a characteristic of the aircraft and presented as function of the yawing moment coefficient in the aircraft technical manuals.

From Fig. 8-14, a sum of the moments about the c.g. of the aircraft results in the following expression:

$$D_{WM}l_{ei} + \sum T_n l_{e_o} = F_t l_t = N = C_n qSb = Yawing\ Moment$$

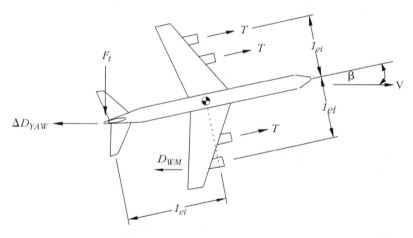

Fig. 8-14 *Forces acting on the aircraft with one engine inoperative.*

In the above equation, l_{ei} is the arm of the inoperative engine, and l_{eo} is the arm of the operating engines; whereas "b" is a representative length of the aircraft, used to nondimensionalize the moment coefficient, much the same way as the wing area "S" is used for the same purpose in all aerodynamic coefficients.

The drag-induced moment of the inoperative engine, and the moment produced by the thrust of the operating engines, must be balanced by the moment that the tail can generate from rudder deflection.

Climb at set thrust

Isolating the moment coefficient, C_n, to express it in terms of the other parameters and Mach number (see Equation 3-8B) yields the following expression:

$$C_n = \frac{\dfrac{D_{WM}l_{e_i}}{\delta} + \dfrac{1}{\delta}\sum Tl_{e_o}}{(1481.35)SbM^2}$$

in which the first term in the numerator, the windmilling drag, is found during flight test, and the results made available in the aircraft manuals. The thrust term is evaluated from knowledge of the aircraft's dynamic state and also the power setting.

Once C_n is calculated, yaw drag can be obtained from the curves available in the aircraft's manuals, where yaw drag is presented as a function of C_n (Fig. 8-15).

From Equation 8-6, we can now calculate the ROC by substituting all the other quantities that are already known.

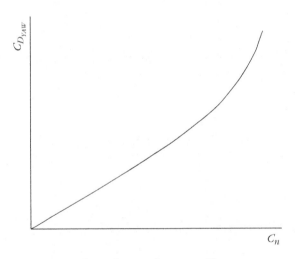

Fig. 8-15 *Plot of yaw-drag coefficient versus yawing-moment coefficient.*

Climb at thrust required

The calculation of climb performance with one or more engines inoperative in a thrust-required mode is somewhat different in that the exact thrust required is not known until the yaw and windmilling drags are known; in turn, yaw drag is not available without knowledge of the thrust produced by the operating engines. This apparent vicious circle can be broken by the process of iteration:

1. Equate lift to weight, and calculate C_L from Equation 3-6.
2. Look up $C_D = f(C_L, M)$ (performance manuals).

3. Calculate drag from Equation 3-7.
4. Set the thrust required, T_{req}, equal to the total drag calculated.
5. Divide T_{req} by the number of operating engines, "n," to obtain the thrust required per engine, T_n.
6. Look up the windmilling drag (performance manuals).
7. Calculate the yawing moment, $N = l_e(T_n + D_{WM})$.
8. Calculate the yawing moment coefficient, C_n, from Equation 3-8.
9. Look up $C_{DYAW} = f(C_n)$ (performance manuals).
10. Calculate D_{YAW} from Equation 3-7.
11. Calculate total drag as the sum of the base drag, D_{WM}, and D_{YAW}.
12. Return to Step 4 and iterate until the resulting value of C_{DYAW} is only marginally larger that the value calculated in the previous iteration (usually 2 iterations suffice).

Example 8-3

A twin jet transport, with the particulars given below, flies at M = 0.3, at 10,000 feet when one of its engines flames out. What will be the thrust required of the remaining engine to maintain the same regime of flight?

Engine moment arm, l_e = 15 ft
Characteristic length (wing span) = 100 ft
Wing area = 1,000 ft^2
Weight = 110,000 lb.
$C_D = 0.07C_L^2 - 0.03C_L + 0.03$
$C_{DWM} = 0.003 - 0.001M$
$C_{DYAW} = C_N^2 + 0.08C_N$

1. First we equate the weight to the lift and calculate C_L:

$$C_L = \frac{110,000}{1481.4\delta SM^2} = \frac{110,000}{(1481.4)(0.6877)(1000)(0.3)^2} = 1.1997$$

2. C_D is given as a function of M:

$$C_D = 0.7(1.997)^2 - 0.03(1.997) + 0.03 = 0.09476$$

3. The drag is now calculated:

$$D = (0.09476)(1481.4)(0.6877)(1000)(0.3)^2 = 8688 \; lb$$

4 & 5. We now set this drag equal to the thrust required, to be produced by the remaining engine.

6. The windmilling drag is calculated next from the expression given. This step is incorporated here, but could have been accomplished earlier:

$$C_{DWM} = 0.003 - 0.001(0.3) = 0.0027$$

$$D_{WM} = 1481.4\delta SM^2 C_{DWM} = (1481.4)(0.6877)(1000)(0.3)^2(0.0027) = 247.6 \; lb$$

7. Next, the yawing moment is calculated:

$$N = l_e(T + D_{WM}) = 15(8688.5 + 247.6) = 134,040 \ in \ lb$$

8. The yawing moment coefficient becomes:

$$C_N = \frac{134,040}{1481.4\delta SM^2 b} = \frac{139,656}{(1481.4)(0.6877)(1000)(0.3)^2(100)} = 0.01462$$

9. The yaw drag coefficient is then calculated from the given formula:

$$C_{D_{YAW}} = C_N^2 + 0.08C_N = (0.01462)^2 + 0.08(0.01462) = 0.00138$$

10. The actual yaw drag can now be calculated:

$$D_{YAW} = (0.00138)(1481.4)(0.6877)(1000)(0.3)^2 = 126.8 \ lb.$$

11. The total drag is then:

$$D_T = D + D_{WM} + D_{YAW} = 8688.5 + 247.6 + 126.8 = 9062.9 \ lb.$$

12. The process is then repeated starting at Step 4, where the thrust required is set equal to the total drag we have just calculated:

$$T = 9062.9 \ lb.$$

6'. The windmilling drag coefficient remains the same at 0.0027, and the corresponding drag 247.6 lb.

7'. The yawing moment now becomes:

$$N = l_e(T + D_{WM}) = 15(9062.9 + 247.6) = 139,656 \ in \ lb.$$

8'. The corresponding moment coefficient is then:

$$C_N = \frac{139,656}{1481.4\delta SM^2 b} = \frac{134,040}{(1481.4)(0.6877)(1000)(0.3)^2(100)} = 0.01523$$

9'. The yaw drag coefficient becomes:

$$C_{DYAW} = C_N^2 + 0.08C_N = (0.01523)^2 + 0.08(0.01523) = 0.00145$$

10'. The corresponding yaw drag is:

$$D_{YAW} = (0.00145)(1481.4)(0.6877)(1000)(0.3)^2 = 133.0 \ lb.$$

11'. And the total drag becomes:

$$D_T = D + D_{WM} + D_{YAW} = 8688.5 + 247.6 + 133.0 = 9069.0 \ lb.$$

Note that the base drag remains the same as calculated in the first set of steps; the only change appears to be the yaw drag.

One more iteration will produce a total drag of 9,069.1 lb., which is close enough to the previous iteration to claim that the process has converged to the final value. Actually, further iterations will produce changes in the total drag only in the second or third decimal places.

In dealing with aircraft with one engine mounted on each wing, the moment arm of the operating and inoperative engines is the same. In aircraft with two engines mounted on each wing, the one-engine-inoperative case requires that the engine-inoperative windmilling drag be balanced by the thrust of the engine in the same position on the other wing. True, there are two other engines that we seem to be neglecting, but their individual moment contributions cancel each other out, thereby obviating the need to include them in the analysis. Nevertheless, for an installation with more than one engine on each wing, a dual engine failure does require that all remaining engines, and their associated moments, be included in the analysis.

Economy climb

Several different climb schedules can be followed in reaching the initial cruise altitude:
- Maximum angle
- Minimum time
- Minimum fuel
- Minimum cost

In the airline environment, three considerations have the most influence on the manner in which aircraft are flown: safety, ATC, and costs. After satisfying the requirements of the first two, the operator will attempt to minimize costs and thereby maximize profit. Minimum-cost climbs are therefore the norm in daily airline operations.

It is important to point out that minimum cost does not necessarily imply minimum fuel. Time is a valuable commodity; it has a value that must be balanced against the value of fuel to arrive at the optimum speed at which to fly. The operator finds this balance and expresses it in terms of cost index (CI), the ratio of time costs to fuel costs, which has units of lb. per hour (lb./hr.). Chapter 10 expands on the concept of CI.

In the regime of flight pertinent to climb and cruise, the faster the aircraft flies, the more fuel it burns. It is up to the individual operator to assess costs accurately enough to determine the relative values of fuel and time and determine the speeds necessary to minimize the cost associated with flying the aircraft.

It is customary to climb at constant CAS/Mach speed combinations, as explained earlier in this chapter, because it presents a relatively simple dynamic situation to both the flight crew and ATC, and because it closely approximates the schedule for a minimum-time climb to initial cruise altitude.

The Mach portion of the CAS/Mach combination is usually determined by the value of Mach to be used once cruise altitude is reached. If the aircraft climbs at cruise Mach, upon reaching cruise altitude there is no need to accelerate (additional fuel), only to pitch over the nose and bring the ROC to zero, while concurrently reducing the power to maintain the desired speed.

The CAS component of the CAS/MACH schedule that results in minimum cost requires more thought: Again, the intent is to minimize the cost of operation by flying a trajectory that results in a compromise between fuel and time costs as established by the operator. Minimizing only the fuel consumed or the time to cruise altitude only addresses part of the requirements.

To assess the consumption of fuel and time for different climb schedules, it is necessary to use similar conditions for all the climb trajectories to be considered. It would be inaccurate to measure the fuel or elapsed time between the beginning and the end of the climb, as the lower speeds will generally result in steeper climbs that take longer to complete. To measure all trajectories on a common basis, we measure both the fuel and time required to go from the beginning of the climb to some arbitrary distance beyond the top of climb (TOC) point, at cruise altitude. This arbitrary distance will be sure to include the point at which the shallowest climb is completed (Fig. 8-16).

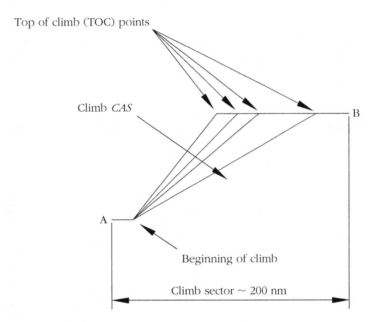

Fig. 8-16 *The effect of CAS on TOC location.*

Measuring time and fuel for the climb sector, from point A to point B, as described, results in plots of fuel and time versus CAS as shown in Figs. 8-17 and 8-18.

It is not surprising that the general shape of a fuel versus CAS plot would be as shown: The higher CAS trajectories are shallower incurring in the double fuel penalty of higher speed and lower altitude. Conversely, Fig. 8-18 seems to defy intuition by indicating that there is a speed above which the climb will take longer to complete. Figure 8-19 contains a plot of both thrust and drag (which could be generalized to thrust required) versus CAS. We have seen that ROC =

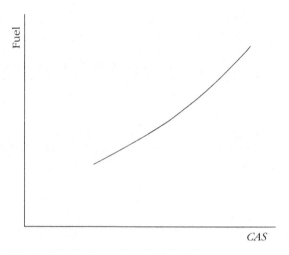

Fig. 8-17 *Fuel used in the climb segment as a function of CAS.*

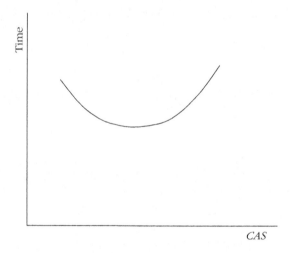

Fig. 8-18 *Time elapsed in the climb segment as a function of CAS.*

$(T - D)/W$, so it is not surprising that there is an optimum CAS where the time to climb will be minimized. Although these curves shift as both altitude and weight change, the concept remains unchanged.

If both time and fuel to reach a common point in cruise are measured for different values of CAS and the same value of Mach, a plot such as the one shown in Fig. 8-20 can be drawn. From it we can determine the minimum fuel CAS and the associated time. We now draw straight lines such as shown in Fig. 8-21, and note that each of the straight lines shown in the plot has a unique

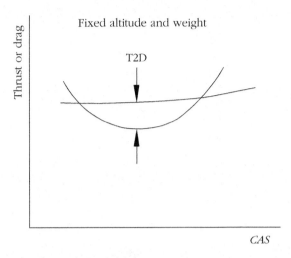

Fig. 8-19 *(T–D) versus CAS.*

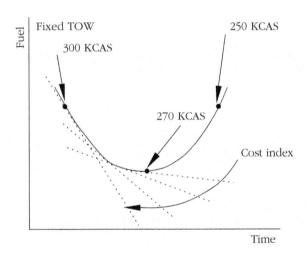

Fig. 8-20 *Cost index is represented by a tangent line to the climb-fuel versus climb-time plot.*

slope in units of pounds of fuel per hour (lb./hr.). Operators, as we mentioned earlier, establish this ratio from knowledge of their cost structure and therefore can choose the line that applies to their particular operation.

The straight lines shown in Fig. 8-21 have the general form

$$y + mx = b$$

where m is the slope of the line (a negative number in this case), and b is the value of the ordinate when the abscissa has a value of zero. From another perspective, we can say that along each line b is a constant representing the total

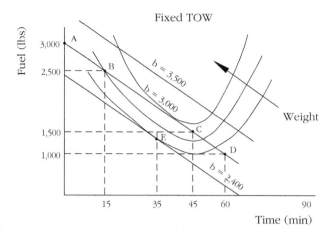

Fig. 8-21 *Climb speed, climb fuel, time to climb, and the effect of weight.*

amount of time and fuel units expended at a ratio of *m*-to-1. A higher value of *b* indicates that more time and fuel are expended, or, more costs incurred. In other words, *b* is the total cost of the operation that results from expending 1 unit of fuel and *m* units of time.

In Fig. 8-21, all the lines are of slope −2,000 (*m* = −2,000); furthermore, all the points along the line for *b* = 3,000, as an example, share the same total cost of 3,000 units, at a ratio of 2,000 units of fuel (lb.) for every unit of time (hr.). Obviously, we would like to operate at the lowest possible cost; we therefore choose to operate at point E, where the total cost is 2,400. Any movement along the operating line of the aircraft (the actual curve) will result in an operation on a straight line of higher *b* value, a line of higher cost.

At higher weights, the aircraft will operate on a different curve, as shown in Fig. 8-21. These curves will have a more pronounced curvature: As the aircraft deviates from the most economical speed, the increase in fuel penalty is more pronounced at the higher weights.

Point E, as can be seen from Fig. 8-21, can be associated with a value of climb CAS, which we proceed to define as best economy climb CAS (ECON CLB CAS) for the particular combination of CLB Mach, TOGW, TOC altitude, and environmental conditions. Mapping all of the possible combinations of these parameters is an extensive undertaking that the manufacturer completes as part of the aircraft development process; the information is provided to the customer via the aircraft manuals, incorporating the information in the FMC, or both. One such presentation might be as shown in Fig. 8-22.

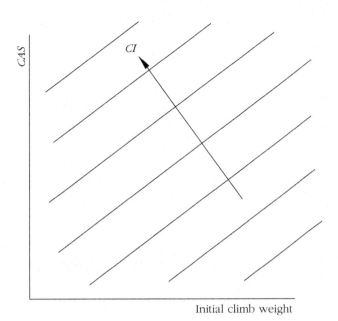

Fig. 8-22 *CAS versus* CI *and initial climb weight.*

9

Cruise

The study of cruise is significant because it is a phase of flight in which additional flexibility is available to the operator to establish the policies by which the aircraft is to be flown in pursuit of a profitable operation. It is important to understand also that ATC often will impose restrictions that must be met by the operator, but whose impact can be minimized by a judicious operation of the aircraft.

In search of more control over the operation of aircraft, the operators of a modern jetliner have at their disposal detailed data and analysis tools that serve them well when the time comes to establish the best speed schedules for their operations. Unfortunately, there are many older jetliners still in service for which the data provided by their manufacturers retains the simplicity and unwarranted assumptions associated with older designs. This chapter will visit the range and endurance formulations, in which the aircraft performance is often expressed in terms of the *thrust specific fuel consumption* (TSFC). More modern aircraft use the concept of specific range instead of TSFC, and the later sections of this chapter are dedicated to their study.

Force-speed diagrams

Chapter 6 discussed thrust available and thrust required as they change with Mach number. On the same graph, it is possible to plot other functions of interest, namely fuel consumption rate (lb./hr.), and *specific range* (*SR*, nam/lb.). The latter will be the subject of more detailed discussion later in this chapter; suffice it to say for now that specific range is a measure of the distance covered per pound of fuel burned, or the mileage of the aircraft. Figure 9-1 shows all these curves as they would appear in a typical presentation.

To include fuel flow in the same plot is a relatively simple process since, as explained in chapter 5, fuel flow depends on the same variables shown in Fig. 9-1, namely, Mach, altitude (represented by the parameter δ), and thrust; for every combination of these parameters, there exists a value of fuel flow.

Considering now that TAS can be expressed as

$$TAS = Ma_0\sqrt{\theta}$$

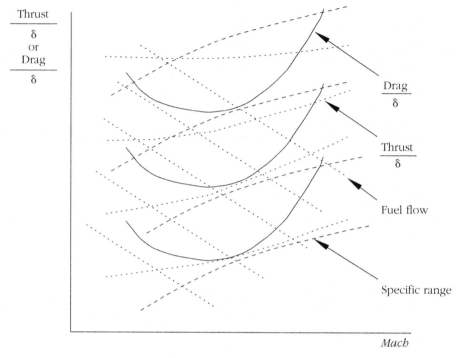

Fig. 9-1 *Drag, thrust, fuel flow, and* SR *versus Mach.*

and that θ is a function of altitude, for every value of Mach along the horizontal axis there corresponds a value of TAS. Knowing the values of fuel flow and TAS, we can form the ratio TAS/ff, which we define as specific range (*SR*), and plot it along with the other parameters in the force-versus-speed plot. The use of *SR* will be apparent later in the chapter.

Speed stability

Consider the plot shown in Fig. 9-2, which is a more detailed look of the thrust and drag curves of Fig. 9-1. Two thrust settings are shown: the higher one with two equilibrium points at A and C, and a lower one with only one equilibrium point at B. At equilibrium condition A, any disturbance that has the effect of reducing the speed of the aircraft along the path marked "a" will result in an excess thrust condition, where the thrust is that corresponding to point A, but the drag has been reduced to some value corresponding to a point between A and B. As a result, the excess thrust will provide the force to accelerate the aircraft back to point A. Likewise, a disturbance resulting in an increase in speed beyond A will result in a thrust deficiency when compared to the increase in drag associated with the higher speed. The resulting imbalance also will cause the aircraft to have the tendency to return to point A. We can expect this behavior for all points to the right of point B, in the region of speed stability.

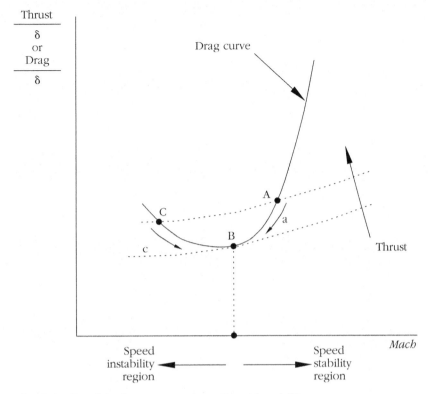

Fig. 9-2 *Graphical representation of speed stability.*

Conversely, if the equilibrium point is at point C, an increase in speed along the path marked "c" will result in a thrust surplus that will tend to accelerate the aircraft even faster. Alternatively, and of more concern, is what happens if the aircraft at point C experiences a disturbance resulting in a speed decrease, with the associated thrust deficiency; in this case, the aircraft will have the tendency to slow down even further, possibly approaching the region of stall. The speed regime to the left of point B is therefore called the *speed instability region*, where aircraft are commonly said to be "behind the power curve." Since aircraft are precisely in this speed region during takeoff and landing, the dynamics exhibited in the speed instability region are of prime concern to those studying wind shear effects.

Maximum endurance

Let us turn our attention now to the fuel flow curves as shown in Fig. 9-3.

The lines of constant fuel flow have a slope such that they become tangent to the drag lines at points such as A and B. Furthermore, the higher the fuel flow line is located in the graph, the higher the value of fuel flow associated with it. Consequently, we conclude that points A and B represent conditions for mini-

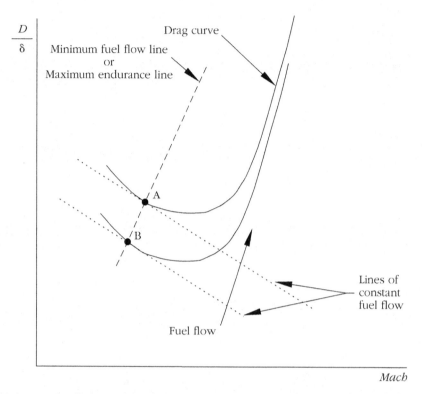

Fig. 9-3 *The maximum endurance schedule is determined by tangency with lines of constant fuel flow.*

mum fuel flow, or maximum endurance, for their corresponding drag curves. A line joining all such points for different drag curves (different weights or altitudes) will constitute the schedule for maximum endurance speeds. Note that due to the slope of the fuel-flow lines of this particular case, the maximum endurance speeds are in the speed instability region.

Maximum range

Even though specific range will be covered in more detail later, we will now show the effect it has in defining the maximum range speeds. Specific range, the ratio of TAS to fuel flow, has units of nautical air miles (NAM, or nam) per pound of fuel, and is a measure of the fuel required to cover some distance. Obviously, the higher *SR*, the more distance covered on a given quantity of fuel.

SR increases downward, as shown in Fig. 9-4, indicating that points C and D represent conditions of maximum range for those particular drag curves. Points A and B represent conditions of minimum drag, and are evidently associated with speeds slower than those for maximum range; therefore, it is wrong to generalize, as is commonly done, that the speed for maximum range is the minimum-drag speed, or maximum lift-over-drag speed.

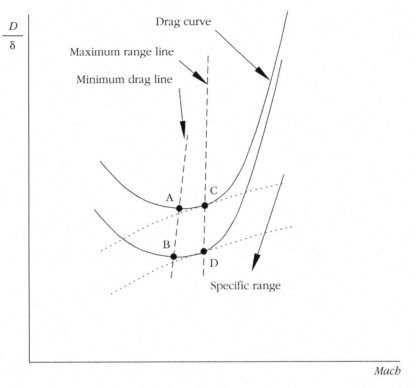

Fig. 9-4 *The maximum range schedule is determined by tangency with lines of constant* SR.

A line joining points of maximum *SR*, such as C and D, is, therefore, the speed schedule for maximum range.

Speed schedules

Figure 9-5 summarizes the speed schedules that have been considered so far in this chapter.

We conclude, then, that the maximum range speed is generally higher than the minimum drag speed, which in turn is higher than the maximum endurance speed. The exact relationship between these speeds, of course, depends on the slopes of the fuel flow and *SR* curves and their tangency points with the drag curves. Each aircraft has its own characteristic set of lines.

Specific range

Specific range (*SR*) is the parameter used to measure the aircraft's mileage in an equilibrium cruise condition. It is measured in nautical air miles (NAM) per pound of fuel. *SR* is formally defined as the ratio of TAS to fuel flow:

$$SR = \frac{TAS}{ff} \tag{9-1}$$

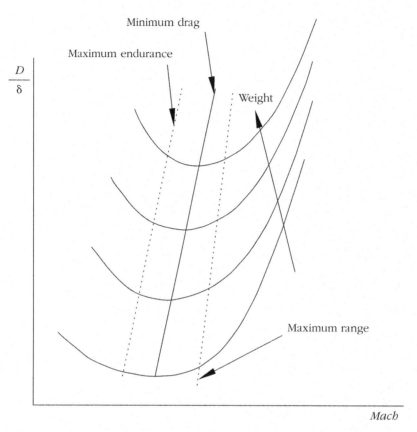

Fig. 9-5 *Speed schedules for maximum endurance, minimum drag, and maximum range.*

where the TAS is measured in NAM per hour, and the fuel flow is measured in pounds per hour.

SR does not include the effect of wind in its definition, as a result of which nautical air miles are used instead of nautical miles. When dealing with *SR*, we are concerned with the ability of the aircraft to travel a distance within the air mass with a given amount of fuel, regardless of the ground covered. *SR* is a measure of the aircraft efficiency in stable cruise and is often used to compare the performance of different aircraft, different flight regimes for the same aircraft, or different configurations of the same aircraft. Wind will naturally have an effect on performance, but is arbitrarily left out when considering SR. We will nevertheless consider the effects of wind later in this chapter.

Figure 9-6 contains a typical plot of *SR* as a function of Mach number and weight. Such plots can be found in the aircraft's performance manuals for several typical cruise altitudes.

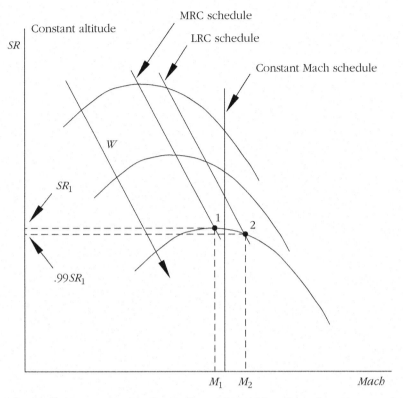

Fig. 9-6 SR *versus Mach and weight with selected cruise-speed schedules.*

Point 1 of Fig. 9-6 corresponds to the highest *SR* value for that aircraft weight; it is called the *maximum range cruise* (MRC) point for that weight and altitude, and M_1 is the corresponding MRC Mach number. When an aircraft is flying at the MRC point for its weight and altitude, we say that it is flying at MRC Mach, or simply MRC, and it is at this value of M that the aircraft will travel the most air miles per pound of fuel consumed; as we have seen, this speed is not generally the minimum drag speed.

The line joining all the MRC points is the MRC speed schedule, and it is the locus of points that will yield the maximum possible air distance per pound of fuel. Naturally, as the aircraft burns fuel and decreases its weight, the corresponding MRC Mach decreases along the MRC speed schedule.

Point 2 in Fig. 9-6 corresponds to a flight condition where the *SR* is 99 percent of the *SR* at MRC. There is no rigorous analytical derivation associated with this percentage; it is an arbitrary definition, a convention followed by the industry. Point 2 is called the *long range cruise* (LRC) point, to which corresponds the LRC Mach at M_2. The LRC speed schedule consists of all the points whose *SR* is 99 percent of MRC and are on the right side of the MRC schedule.

Example 9-1

The accompanying figure shows *SR* curves for three altitudes. What are the MRC and LRC values for these altitudes?

At FL350, the peak of the curve occurs at $M = 0.75$, with $SR = 0.0708$; this is MRC Mach. If we take 99 percent of the *SR* value, or 0.0700 approximately, we see that it corresponds to $M = 0.72$ or $M = 0.78$, depending on which side of the maximum value we look at. But LRC is defined on the fast side of the curve, namely $M = 0.78$.

A similar process yields the following results:

	MRC (SR/M)	LRC (SR/M)
FL350	0.07080/0.75	0.07000/0.78
FL370	0.07247/0.76	0.07170/0.78
FL390	0.05670/0.77	0.05610/0.79

The MRC points are identified with the small circle, and the LRC points are identified with a star.

Note that the flight regime for best fuel mileage is FL370. Flying above or below that flight level incurs a penalty; the penalty is substantial at FL390, even though it is generally believed that the higher the altitude, the lower the fuel consumption. This might be true if the speed is kept constant—fuel flow might be lower—but the speed necessary to maintain equilibrium aerodynamic conditions is higher.

The LRC schedule has been widely adopted by the industry as the speed at which to fly to derive the most benefit from the available fuel; it is an inaccurate label, but one that has stuck throughout the industry.

As stated before, Fig. 9-6 can be plotted for several altitudes. If we have several plots for different altitudes, we can study a particular speed schedule in all of them, say LRC (or MRC, or constant Mach). From all the possible points along the LRC schedule, we also elect to work with a particular value of weight, say, the weight corresponding to point 2 in Fig. 9-6. Reading the value of *SR* associated with each one of these points, in each of the plots made for several altitudes, we can derive another plot, one of *SR* versus altitude, as shown in Fig. 9-7.

Clearly, for every value of weight, the point of maximum *SR* represents the altitude at which maximum range can be achieved for the chosen weight and speed schedule. This altitude is called the *optimum altitude for maximum range*, or max range opt alt.

Notice that we pursued this exercise with an LRC schedule; we could just as well have done the same with the MRC or constant Mach schedules, thus deriving a max range opt alt for them also.

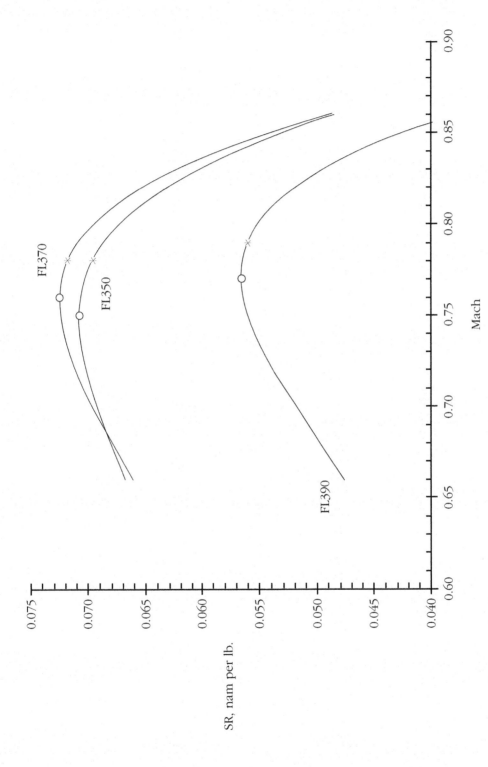

Figure for example 9-1.

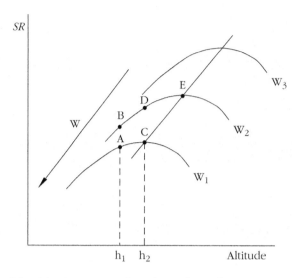

Fig. 9-7 SR *versus altitude and weight.*

Temperature effects

The effect of ambient temperature on *SR* can be appreciated by looking at the definition of *SR* and its components—in the case of older formulations, in terms of SFC:

$$SR = \frac{TAS}{ff} = \frac{a_0\sqrt{\theta}M}{SFC\left(\dfrac{F_n}{\sqrt{\theta}}\right)\delta\sqrt{\theta}} = \frac{a_0\sqrt{\theta}M}{SFC\,F_n} \propto \sqrt{\theta} = \sqrt{\frac{T}{T_{amb}}}$$

In the more current presentations, fuel flow is given as corrected fuel flow, W_{fc},

$$W_{fc} = \frac{W_f}{\delta_T\theta_T^x}$$

as we saw in chapter 5. Again, *x* is an exponent that depends on the power-plant; its value is usually around 0.5.

SR then becomes

$$SR = \frac{TAS}{W_f} = \frac{Ma_0\sqrt{\theta}}{W_{fc}\delta_T\theta_T^x}$$

Referring to Equations 1-10 and 1-11, repeated here for convenience,

$$\delta_T = \delta\,(1 + 0.2M^2)^{3.5} \tag{1-10}$$

$$\theta_T = \theta(1 + 0.2M^2) \tag{1-11}$$

we can rewrite W_f as

$$W_f = W_{fc}\delta\theta^x(1 + 0.2M^2)^{3.5 + x} \tag{9-2}$$

Now, rewrite the standard temperature SR as

$$SR_{st} = \frac{TAS}{W_f} = \frac{a_0 M \sqrt{\theta_{st}}}{W_{fc}\delta\theta_{st}^x(1 + 0.2M^2)^{3.5 + x}} \tag{9-3}$$

and, for nonstandard conditions,

$$SR_{ns} = \frac{TAS}{W_f} = \frac{a_0 M \sqrt{\theta_{ns}}}{W_{fc}\delta\theta_{ns}^x(1 + 0.2M^2)^{3.5 + x}} \tag{9-4}$$

Dividing Equation 9-4 by Equation 9-3 for constant Mach and altitude,

$$\frac{SR_{ns}}{SR_{st}} = \frac{\dfrac{\sqrt{\theta_{ns}}}{\theta_{ns}^x}}{\dfrac{\sqrt{\theta_{st}}}{\theta_{st}^x}} = \frac{\theta_{ns}^{0.5 - x}}{\theta_{st}^{0.5 - x}} = \left(\frac{T_{ns}}{T_{st}}\right)^{0.5 - x}$$

which allows us to write, for a nonstandard day,

$$SR_{ns} = SR_{st}\left(\frac{T_{ns}}{T_{st}}\right)^{0.5 - x} \tag{9-5}$$

Equation 9-5 allows the ready evaluation of SR for nonstandard conditions based on the data available for standard conditions.

Figure 9-8 is a graph of SR in which thrust limiting curves have been included. These curves represent the limitations imposed by the engines when operating at a rating (usually MCR in cruise). Accordingly, if the temperature is such that at the altitude for which the plot was made, the ambient temperature is ISA+ΔT, the MCR rating available at that temperature would limit the operation of the aircraft to all points to the left of the ISA + ΔT line, with $M = M_1$ being the highest Mach achievable. A similar argument applies to ISA and ISA–ΔT conditions, for which the limiting values of M are M_2 and M_3 respectively.

As long as the aircraft is operated off the rating limit line, and to the left of it, Equation 9-5 can be used to calculate SR for nonstandard conditions from the data available for standard conditions.

If we conduct this exercise and plot the resulting curves, the result will be as shown in Fig. 9-9. When the aircraft is operated precisely at the intersection of the SR and MCR curves (at the MCR rating), for any change in temperature, there will be one change in SR due to the shift in the SR curve and another change due to the shift in the rating curve. Under this double effect, in temperatures hotter than standard, the power must be reduced in accordance with the rating schedule, thereby making the operation more efficient (point 2). The opposite occurs in colder temperatures, and the aircraft is operated at increased power and reduced efficiency (point 3).

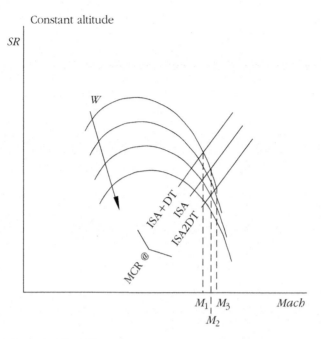

Fig. 9-8 *The effect of temperature on the maximum* M *attainable at MCR.*

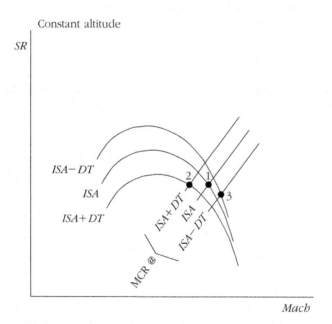

Fig. 9-9 *Effect of temperature on* SR *and its effect on maximum* M *attainable at MCR.*

Wind and altitude effects

While in the cruise portion of the flight, the crew often finds it necessary to change the cruise altitude, either to find smoother weather, or to take advantage of more favorable winds. Whatever the circumstances, it is important for the crew to assess the impact of the wind at the new altitude on the expected range of the aircraft.

The following effects should be considered when changing cruise altitude:
- SR decreases with headwind, but the effect is minimized by increasing the cruise speed by an amount that depends on the magnitude of the wind.
- An altitude increase requires the expenditure of fuel.
- An altitude decrease can be accomplished at the expenditure of potential energy, minimizing the consumption of fuel.

Any change in altitude will bring into play one or more of the above effects to a varying degree, thus requiring wise energy management to optimize the performance of the aircraft.

If we take the definition of SR as presented in Equation 9-1 and consider the definition of ground speed:

$$GS = TAS + V_W$$

where the value of the wind velocity, V_W, is negative for headwinds and positive for tailwinds, we have the ingredients for an expression of SR in terms of the ground speed,

$$SR = \frac{GS - V_W}{ff} \tag{9-6}$$

or

$$SR = \frac{GS}{ff} - \frac{V_W}{ff} \tag{9-7}$$

The first term in the right-hand side of Equation 9-7 can be considered the ground-based SR, SR_G, measured in nautical miles (NM), not nautical air miles (NAM). Thus,

$$SR_G = SR + \frac{V_W}{ff} \tag{9-8}$$

Substituting Equation 9-1 into Equation 9-8 to eliminate ff yields

$$SR_G = SR + \frac{V_W SR}{TAS}$$

or

$$SR_G = SR \left(1 + \frac{V_W}{TAS} \right) \tag{9-9}$$

Substituting now the definition of Mach Number,

$$SR_G = SR \left(1 + \frac{V_W}{a_0 M \sqrt{\theta}} \right) \tag{9-10}$$

We call the quantity within the parentheses of Equation 9-10 the *wind shift function* (WSF). For zero wind, $WSF = 1$, and $SR_G = SR$; otherwise, when $V_W \neq 0$, the *WSF* shifts the *SR* curve, reflecting the effects of wind on the range capability of the aircraft.

In Fig. 9-10, we can appreciate the effect of different intensities of headwind on SR_G curves for FL250 and FL390. Similar but opposite effects would ensue for tailwinds.

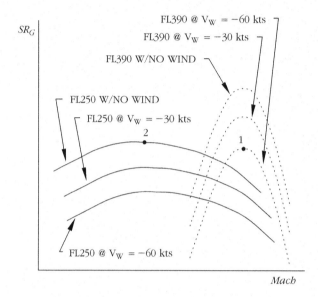

Fig. 9-10 *Effect of wind on* SR_G.

This presentation allows the study of the effects of changes in altitude in search of better wind conditions; the goal is to maximize the gain (minimize the loss) in SR_G with any altitude change. For example, flying at FL390, at optimal Mach, in a 60-knot headwind (point 1) incurs a value of SR_G just below the value corresponding to a similar flight condition at FL250 with no wind (point 2); therefore, a change to FL250 is beneficial because it will improve the ground-relative range of the aircraft. Note that in this treatment we are not concerned with the possibility of flying more slowly and therefore extending the time of flight slightly. Our concern here is restricted to improving the range of the aircraft in the presence of winds and not with the time it takes to complete the trip. When time is important, we must weigh its value with the value of fuel, something we will consider in the chapter on economy cruise.

Example 9-2

The table for Example 9-2 has been created from data for a 200,000-pound aircraft flying at different altitudes in the presence of different wind conditions. The

Table for example 9-2

Specific range

	No wind			100 Knots of headwind			0 kts	250 kts	2100 kts
M	FL350	FL370	FL390	FL350	FL370	FL390	FL350	FL370	FL390
0.720	0.065194	0.064260	0.062595	0.049487	0.048700	0.047438	0.065194	0.056480	0.047438
0.725	0.065470	0.064657	0.063064	0.049805	0.049109	0.047899	0.065470	0.056883	0.047889
0.730	0.065716	0.065041	0.063527	0.050100	0.049508	0.048355	0.065716	0.057275	0.048355
0.735	0.065928	0.065408	0.063978	0.050367	0.049894	0.048803	0.065928	0.057651	0.048803
0.740	0.066100	0.065751	0.064410	0.050605	0.050261	0.049236	0.066100	0.025006	0.049236
0.745	0.066320	0.066064	0.064818	0.050809	0.050604	0.049650	0.066230	0.058334	0.049650
0.750	0.066313	0.066339	0.065192	0.050974	0.050919	0.050038	0.066313	0.058629	0.050038
0.755	**0.066342**	0.066570	0.065524	0.051099	0.051198	0.050393	**0.066342**	0.058884	0.050393
0.760	0.066314	0.066747	0.065803	0.051177	0.051436	0.050708	0.066314	0.059091	0.050708
0.765	0.066222	0.066861	0.066019	**0.051205**	0.051624	0.050974	0.066222	0.059243	0.050974
0.770	0.066062	**0.066903**	0.066161	0.051179	0.051179	0.051181	0.066062	0.059329	0.051181
0.775	0.065827	0.066863	**0.066214**	0.051093	**0.051822**	0.051319	0.065827	**0.059343**	0.051319
0.780	0.065512	0.066729	0.066167	0.050942	0.051815	**0.051378**	0.065512	0.059272	**0.051378**
0.785	0.065109	0.066491	0.066004	0.050721	0.051724	0.051345	0.065109	0.059108	0.051345
0.790	0.064614	0.066136	0.065709	0.050425	0.051541	0.051209	0.064614	0.058839	0.051209
0.795	0.064018	0.065652	0.065267	0.050048	0.051255	0.050955	0.064018	0.058453	0.050955
0.800	0.063314	0.065024	0.064660	0.049585	0.050854	0.050570	0.063314	0.057939	0.050570

columns for the no-wind condition come directly from the performance documents for the aircraft for FL350, FL370, and FL390. The other columns were created using Equation 9-10. Highlighted in bold letters in all columns are the MRC conditions for each altitude/wind combination.

Under the no-wind condition, the altitude that yields the best range performance is FL370 at $M = 0.770$. With a 100-knot headwind at all altitudes, the best range is still obtained at FL370, but at a higher Mach (0.775); in other words, the curves have shifted to the right, and, of course, to lower values of SR.

Alternatively, if the wind conditions are as shown in the last three columns, the best range performance occurs at FL350 and M = 0.755.

Although the concept of optimum altitude can be applied to SR_G, it loses value when we consider that the result would be the altitude that yields the maximum range in the presence of constant wind for all altitudes. Rarely will the wind be the same for all altitudes under consideration. A more likely condition is that the wind will change as altitude changes, and the crew or flight-plan designer are tasked with the job of finding the altitude/wind combination that will yield the best range.

In this section, we have used specific range to define the best speed for fuel economy in the presence of winds and also as a tool in prescribing a policy for wind/altitude trades. To achieve the same results, aircraft and flight computer manufacturers prefer to speak of minimizing flight cost:

$$Minimize\left(\frac{ff}{V_g}\right)$$

rather than maximizing SR. The two processes are analogous. Any confusion about their relationship is allayed by noticing that, through the use of Equation 9-1, the above expression for cost is transformed easily into Equation 9-9. More directly, cost as expressed in the above equation is the inverse of SR_G; therefore, maximizing one is the same as minimizing the other.

In the airline environment, it is more convenient to use the SR formulation simply because SR is abundantly presented, usually in the performance engineering manual. By using this presentation directly, the performance engineer avoids the need to calculate fuel flow from tables that invariably present it as corrected fuel flow, W_{fc}. If W_{fc} is to be used, first it must be multiplied by δ and θ^x, where the exponent x is dependent on the engine installation. Furthermore, a more complicated situation is presented if fuel flow has been corrected using δ_t, in which case a slightly more involved calculation involving Mach number must be used.

Step climb

From Fig. 9-7 it is evident that, generally, the best flight path is one whose altitude continuously increases as fuel is burned off; such a flight profile is called a *cruise climb.* Unfortunately, such a mode of operation is rarely allowed by the controlling authorities. More frequently an aircraft will be allowed to change its cruise altitude by a fixed increment, for example from FL350 to FL390. A common practice is, then, to straddle the continuously rising optimum altitude with steps that are compatible with the limitations imposed by ATC.

Under these conditions, the flight would begin its cruise segment 2,000 feet above optimum altitude; eventually, because of fuel burn off, the optimum altitude will increase to the cruise altitude being flown and continue by the same process. When the optimum altitude has increased to an altitude that is 2,000 feet above the aircraft cruise altitude, the crew would climb to 2,000 feet above it. This process would be continued until the end of the cruise segment. For example, suppose the optimum altitude for an aircraft at its initial cruise weight is 34,000 feet; knowing this, the crew elects to climb to 36,000 and initiate the cruise segment there. After some time, the optimum altitude has risen to 36,000 feet and therefore matched the aircraft initial-cruise altitude; at this point, the crew does nothing, maintaining the same cruise altitude of 34,000 feet. Later still, enough fuel has been burned that the optimum altitude has attained a value of 38,000 feet; now the crew would climb to 40,000 and maintain that altitude until the process is repeated again.

In addition to the limitations imposed by ATC, there are two other considerations that impact step climbing: maximum altitude and the top of descent point (TOD). The first one is obvious because the aircraft cannot climb past its current maximum altitude. The second consideration stems from the need to assess if the savings realized by stepping to a higher altitude will be negated by a descent that might have to be initiated shortly thereafter, so that the descent path will end at the intended destination.

Short trip optimum altitude (STOA)

Throughout our discussion, optimum altitude has not been influenced by conditions at either origin or destination, or by the conditions that prevail during the climb and descent. Optimum altitude is therefore a parameter whose value depends only on current conditions.

If the distance between origin and destination is sufficiently short, our concept of optimum altitude is not appropriate. In a relatively short trip, the fuel and time consumed during climb and descent represent a large proportion of the entire flight; consequently, the cruise speed and altitude that would yield the best results for a longer trip must now be adjusted. For example, an aircraft weighing 110,000 lb. might have a conventional long-trip optimum altitude of 35,000 feet; but if the same aircraft is to make a flight between two cities that are 200 nautical miles apart, the amount of fuel required to climb to FL350, cruise, and then descend to the destination might negate any savings accrued by flying at the proposed optimum altitude of 35,000 feet. Indeed, the distance required just to climb to 35,000 feet and then descend might very well be longer than 200 nautical miles. There is a need, therefore, to *distinguish between long trip optimum altitude* (LTOA) and *short trip optimum altitude* (STOA).

We have seen that LTOA depends greatly on the speed schedule followed in cruise, be it LRC, MRC, constant Mach, ECON, or the like. Conversely, STOA, by virtue of the short duration of the cruise portion of the flight, does not depend on cruise parameters very much; instead, it is heavily dependent on the conditions in which the climb and the descent take place, and also on the conditions prevailing at origin and destination airports.

The derivation of STOA is empirical because the parameters on which it depends are not easily convoluted into a mathematical relationship. In general, STOA will depend on the following parameters:

- Takeoff weight. As expected, the heavier the aircraft at takeoff, the lower the STOA. A large quantity of fuel would be burned lifting the aircraft to higher altitudes.
- Climb and descent power settings. The power levels will determine not only the fuel burn rate but also the time spent climbing and descending.
- Origin and destination altitudes. Because the climb and descent segments have gained importance, it is not surprising that the total distance climbed and descended will be important, and the altitudes of the origin and destination airports are therefore relevant.
- Ambient temperature (ISA deviation). The prevailing ambient temperature at all altitudes will have an effect on the power available and the rate of fuel consumption.
- Climb and descent wind gradients. From our study of climb (chapter 8), we know that the climb angle depends not on the wind, but the wind gradient; furthermore, it is the climb angle or gradient that determines the relationship between the altitude attained and the horizontal distance covered in climbing (similarly with the descent).
- Distance between origin and destination. Obviously the distance to be covered by the flight has an impact on the optimum conditions for that flight: the longer the distance, the more room there is to climb to a higher altitude.
- Desired minimum cruise time. Operators may wish to establish a minimum cruise time to allow the crew ample time to accomplish the required procedures.

Through extensive flight test, simulations, and analyses, the manufacturer establishes the STOA as a function of all the above parameters and presents them in the aircraft performance manuals and the FMC.

Integrated range

From the basic concept of *SR*, as shown in Fig. 9-6, we can arbitrarily choose any speed schedule, say LRC, and cross-plot *SR* as a function of weight, resulting in a plot such as shown in Fig. 9-11. This new version of SR we will call SR_W. It will always decrease with increasing weight.

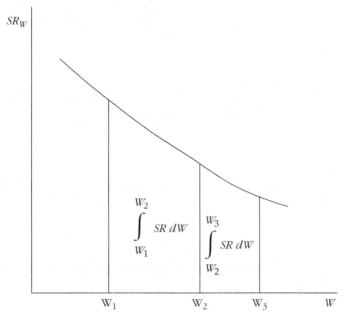

Fig. 9-11 SR *versus weight for a specific speed schedule.*

From the definition of SR, we know that it can be interpreted as the change in distance x flown by consuming a certain amount of fuel:

$$SR_W = \frac{dx}{dW}$$

which, upon solving for dx and integrating, becomes

$$\Delta X_{1,2} = \int_{W_1}^{W_2} SR_W dW, \qquad W_1 < W_2 \qquad (9\text{-}11)$$

The integral in Equation 9-11 depends on the SR curve chosen (for a given altitude and speed schedule) and the limits of integration (the initial and final weight of the aircraft). Note that, for the purposes of integration in Equation 9-11, the initial weight is W_1, and the final weight is W_2, whereas in the real world W_1, the lower weight is necessarily the final weight; in other words, in the process of integration, the roles of W_1 and W_2 are reversed from the order these weights have in the scale of time.

Suppose we have selected an SR curve (FL350 @ M=0.8), and an initial weight, W_1; the integral in Equation 9-11 now depends only on the final weight, W_2. We can therefore elaborate a table as follows:

Table 9-1
Integrated range

WEIGHT	X
W_1	0
W_2	X_1
W_3	X_2
–	–
–	–
–	–
W_n	X_{n-1}

Now let's investigate what happens when we consider a weight W_3 such that $W_3 < W_2 < W_1$:

$$\Delta X_{1,3} = \int_{W_1}^{W_3} SR_W dW = \int_{W_1}^{W_2} SR_W dW + \int_{W_2}^{W_3} SR_W dW$$

or

$$\int_{W_2}^{W_3} SR_W dW = \int_{W_1}^{W_3} SR_W dW - \int_{W_1}^{W_2} SR_W dW \tag{9-12}$$

which implies that

$$\Delta X_{2,3} = \Delta X_{1,3} - \Delta X_{1,2}$$

or, in other words, the distances resulting from evaluating integrals such as Equation 9-11 can be added and subtracted for corresponding weight changes.

With this knowledge, we can now enter Table 9-1 and calculate the fuel required to fly at the chosen speed schedule and altitude from point A to point B starting at a given weight W_A:

Table 9-2
Integrated range

WEIGHT	X
–	–
–	–
–	–
W_A	X_A
–	–
–	–
–	–
W_B	X_B

We know W_A and the quantity $(X_B - X_A)$, the distance to be flown. Entering the table with W_A, we look up X_A. To X_A, we add the distance to be flown and obtain X_B, and then look up the corresponding W_B. The difference $(W_A - W_B)$ will be the weight of the fuel required to fly the prescribed distance.

Integrated time

In a manner similar to integrated range, we can develop a similar argument for the time required to consume a given quantity of fuel:

$$SR_W = \frac{TAS}{ff} = \frac{a_0 M \sqrt{\theta}}{ff} \Rightarrow ff = \frac{a_0 M \sqrt{\theta}}{SR}$$

We also know that fuel flow can be considered the time rate of change of aircraft weight:

$$ff = \frac{dW}{dt}$$

After combining the above two expressions, solving for the time differential and then integrating, the resulting expression is

$$\Delta t = - \int_{W_1}^{W_2} \frac{SR_W dW}{a_0 M \sqrt{\theta}} \tag{9-13}$$

Here again we have chosen an SR_W for a fixed altitude and speed schedule, and therefore can no longer assume that Mach will be constant, whereas θ and a_0 will; so,

$$\Delta t = - \frac{1}{a_0 \sqrt{\theta}} \int_{W_1}^{W_2} \frac{SR_W dW}{M} \tag{9-14}$$

From Equation 9-14 we can develop an integrated time table similar to Table 9-1.

Point of no return (PNR)

The point of no return is the point between origin and destination at which the fuel burn to the destination equals the fuel burn back to the origin; accordingly, the PNR can also be called the *equal fuel point* (EFP).

We do not specify in the definition of PNR the regime of flight under which the outbound and return trips are to be flown; there could and probably will be differences in altitude, wind, and perhaps speed.

Let the outbound incremental fuel burn, from the PNR to the destination, be expressed as

$$dW_{out} = -ff_{out}dt \qquad (9\text{-}15A)$$

and for the inbound leg

$$dW_{in} = -ff_{in}dt \qquad (9\text{-}15B)$$

From the definition of SR (Equation 9-1), we can solve for the fuel flow, ff, and substitute into Equations 9-15 to obtain

$$dW_{out} = -\left(\frac{TAS}{SR}\right)_{out} dt \qquad (9\text{-}16A)$$

$$dW_{in} = -\left(\frac{TAS}{SR}\right)_{in} dt \qquad (9\text{-}16B)$$

We also know that the incremental distance dx equals TAS*dt; therefore,

$$dW_{out} = -\frac{dx}{SR_{out}} \qquad (9\text{-}17A)$$

$$dW_{in} = -\frac{dx}{SR_{in}} \qquad (9\text{-}17B)$$

Integrating Equations 9-17,

$$W_{out} = -\int_{x_{PNR}}^{x_{dest}} \frac{dx}{SR_{out}} \qquad (9\text{-}18A)$$

$$W_{in} = -\int_{x_{PNR}}^{x_{orig}} \frac{dx}{SR_{in}} \qquad (9\text{-}18B)$$

By definition, the PNR is the point, x_{PNR}, at which W_{out} equals W_{in}, or,

$$\int_{x_{PNR}}^{x_{dest}} \frac{dx}{SR_{out}} = \int_{x_{PNR}}^{x_{orig}} \frac{dx}{SR_{in}} \qquad (9\text{-}19)$$

For cruise at constant outbound and inbound altitudes (although generally different from each other), Equation 9-19 can be rewritten as

$$a_0 \sqrt{\theta_{out}} \int_{x_{PNR}}^{x_{dest}} \frac{Mdt}{SR_{out}} = a_0 \sqrt{\theta_{in}} \int_{x_{PNR}}^{x_{orig}} \frac{Mdt}{SR_{in}} \qquad (9\text{-}20)$$

With a detailed description of SR, such as presented in the aircraft performance manuals, Equations 9-19 or 9-20 can be integrated and x_{PNR} found by iteration. If winds are present, it will be necessary to use the function SR_G already discussed.

Note that

$$dW = -\frac{dx}{SR} \Rightarrow SR = -\frac{dx}{dW}$$

which is the expression we used in developing the integrated-range concept. There is, therefore, an alternate method of solution of the PNR problem, by using the integrated range tables available for the aircraft. The solution is achieved by a process of iteration that would begin with an estimate of the fuel burn to the PNR (for example, half the burn from origin to destination).

Tankering

Tankering is the practice of carrying more fuel than is required for the planned flight for the purpose of avoiding to load fuel at the destination airport. This is usually done when the price differential between the fuel at destination and the fuel at the origin warrants the additional weight of the aircraft load.

To assess the desirability of tankering fuel, it is necessary, but not sufficient, to calculate the additional fuel burn associated with the higher weight aircraft. To this end, we turn to the integrated range chart for the desired speed schedule.

To fly from a point 1 to another point 2, a total distance of $x_{1,2}$, the fuel consumed is $W_1 - W_2$. From the formulation of integrated range, we know this can be expressed as

$$x_{1,2} = \int_{W_1}^{W_2} SRdW$$

an integral that is represented by areas A_1 and A_2 combined, as shown in Fig. 9-12. We can also split the integral into two integrals,

$$x_{1,2} = \int_{W_.}^{W'_2} SRdW + \int_{W'}^{W_2} SRdW$$

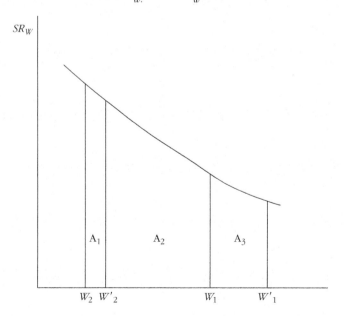

Fig. 9-12 *Breakdown of tankering integral.*

Now, let us arbitrarily choose to tanker a quantity of fuel, W_t, equal to the weight difference, $(W_2 - W_2')$; the aircraft, therefore, will reach point 2 at the higher weight W_2'. By so doing, we have depleted the required area under the SR curve by an amount equal to A_1, which has to be made up if the aircraft is to complete the distance $x_{1,2}$ without using any of the tankered fuel (remember that the areas under the curve represent distance covered). To make up for the lost area (distance), it is necessary to add area A_3, equal to A_1, indicating a higher weight at the beginning, W_1'. Note that because the slope of the curve at the lighter weights is steeper than at lower weights, and because the curve is farther from the weight axis, the weight difference necessary to encompass the same area under the curve is larger at the higher weights than at the lower weights. In other words, to tanker any amount of fuel, it is necessary to burn additional fuel, as expected. The penalty for tankering the quantity of fuel $W_t = (W_2 - W_2')$ is

$$W_p = (W_1' - W_1) - (W_2' - W_2)$$

W_p is therefore the additional fuel required to tanker W_t, which in turn implies that the total fuel burn is $W_1 - W_2 + W_p$, or $(W_1' - W_2')$.

Range and endurance formulations

Many references present range and endurance by deriving expressions for them based on the concept of *thrust-specific fuel consumption* (TSFC). These formulas are generally based on one or more of the following assumptions, commonly used in conjunction with the data presented for older aircraft:

- TSFC is used by the aircraft manufacturer to measure fuel consumption.
- TSFC and the lift-over-drag ratio (L/D) remain essentially constant during cruise.
- Fuel flow is a linear function of thrust.

In engineering, as in many other disciplines, some simplifying assumptions are sometimes necessary to facilitate the solution of practical problems, and the above assumptions serve that purpose well.

Nevertheless, with the increasingly available computers, both for engineering and onboard use, some of these assumptions become unnecessary and sometimes even counterproductive. Many aircraft manuals still present fuel efficiency in terms of TSFC, yet the trend is away from this practice toward a more detailed presentation, where fuel flow is expressed in terms of altitude, Mach, and thrust.

The second and third assumptions are based on a smooth and linear variation of fuel flow with thrust. This smooth linear behavior is not true in general; nonlinear behavior is often introduced by effects of viscosity, temperature, and altitude. In addition, the modern jet transport has complex schedules of engine bleed valve switching that often introduce discontinuities in the fuel flow.

Endurance

It is intuitive that maximum endurance occurs when the fuel flow is at a minimum. Furthermore, traditionally, fuel flow has been calculated as the result of multiplying TSFC by thrust:

$$ff = TSFC \times F_n$$

If we now assume that thrust equals drag in cruise, we conclude finally that, for a constant TSFC, minimum fuel flow occurs at the minimum drag speed.

Consider fuel flow as the negative rate of change of aircraft weight; it could then be written as

$$ff = -\frac{dW}{dt} = TSFC \times F_n$$

This is an expression that can be solved for the differential of time,

$$dt = -\frac{dW}{TSFC \times F_n}$$

which, when integrated, yields the endurance, E,

$$E = -\int_{W_0}^{W_1} \frac{dW}{TSFC \times F_n} = \int_{W_1}^{W_0} \frac{dW}{TSFC \times F_n}$$

where W_0 is the initial weight and W_1 is the final weight.

Assuming now that thrust equals drag, and lift equals weight, we can multiply and divide by lift to obtain

$$E = \int_{W_1}^{W_0} \frac{L}{D} \frac{1}{TSFC} \frac{dW}{W} \qquad (9\text{-}22)$$

Substituting Equations 3-6 and 3-7 and assuming constant TSFC and L/D, Equation 9-22 reduces to

$$E = \frac{C_L}{C_D} \frac{1}{TSFC} \ln \frac{W_0}{W_1} \qquad (9\text{-}23)$$

From Equation 9-23, we conclude that maximum endurance will be obtained when either the TSFC is minimized, the aircraft flies at maximum L/D, or the amount of fuel at the beginning of the cruise segment is maximum. Since we have stated that TSFC does not necessarily remain constant during cruise, a more accurate estimation of the endurance capability will be obtained by studying Equation 9-22. Unfortunately, the integral in that equation cannot be solved without detailed knowledge of the variation of the integrand as the weight of the aircraft changes.

Range

Velocity is defined as the rate of change of distance:

$$V = \frac{ds}{dt}$$

Solving for the differential ds,

$$ds = Vdt = -V\frac{dW}{TSFC \times F_n}$$

Integrating to obtain the range, R,

$$R = -\int_{W_0}^{W_1} \frac{VdW}{TSFC \times F_n} = \int_{W_1}^{W_0} \frac{VdW}{TSFC \times F_n}$$

Solving for the velocity "V" in Equation 3-6A, and substituting in the last equation,

$$R = \int_{W_1}^{W_0} \sqrt{\frac{2}{\rho S}} \frac{\sqrt{C_L}}{C_D} \frac{1}{TSFC} \frac{dW}{\sqrt{W}} \tag{9-24}$$

Assuming now a constant C_L/C_D, ρ, and TSFC the above expression can be readily integrated,

$$R = 2\sqrt{\frac{2}{\rho S}} \frac{1}{TSFC} \frac{\sqrt{C_L}}{C_D} \left(\sqrt{W_0} - \sqrt{W_1} \right) \tag{9-25}$$

From Equation 9-25, we see that the range will be maximized if the TSFC is minimized, the flight is conducted at maximum $\sqrt{C_L}/C_D$ or high altitude (low ρ), or the initial fuel is maximized.

As with endurance, more accurate results will be obtained if Equation 9-24 is used.

If we elect to use Mach instead of airspeed in Equation 9-25, the resulting expression is:

$$R = a_0 \int_{W_1}^{W_0} \frac{\sqrt{\theta}}{TSFC} \frac{M}{W} \frac{L}{D} dW$$

where the quantity $(ML)/(WD)$ is called the *range factor*, and, like TSFC, is sometimes assumed to remain unchanged during cruise.

If all quantities in this last equation remain constant as fuel is burned, the integration can be carried out to yield

$$R = \left(\frac{a_0\sqrt{\theta}}{TSFC} \right) \left(M\frac{L}{D} \right) \ln \left(\frac{W_0}{W_1} \right) \tag{9-26}$$

Again, in most circumstances neither the range factor nor TSFC remain constant; therefore, their use in calculating range or endurance is limited to highlighting the dependence between the different variables, in other words, for qualitative work.

10

Economy cruise

Costs are a paramount concern to a commercial carrier. In the competitive commercial airline environment, the difference between the costs of one airline and those of its competitor might decide the profitability of the company and, indeed, perhaps even its longevity as a viable competitor.

It is readily recognized that the energy required to move anywhere from 50 to 400 tons of aircraft and payload through the air at high speed and altitude represents an enormous quantity of fuel. To address this expense, the modern jetliner is designed to minimize fuel consumption by incorporating the latest developments in aerodynamics, propulsion, and avionics. It is up to the operator to capitalize on these technologies by making use of the aircraft in the most efficient manner, consistent with the requirements imposed by the market to be served.

If fuel consumption were the only concern, flying at MRC speeds would guarantee an efficient operation. Unfortunately, higher speeds are sometimes necessary to accommodate schedules and other similar time requirements. This points to the importance of another resource, whose value is not always easy to quantify: time.

The value of time can now be incorporated effectively into the calculations leading to most efficient speeds thanks to the advent of onboard computers such as flight management systems (FMS).

As an illustration of the relative value of time and fuel consider the following two operators:

The overnight package courier has a need to bring all inbound flights to its distribution hub on a very tight schedule, which will allow it to dispatch the outgoing flights on time to deliver their cargo by early morning. The aircraft therefore will fly as fast as necessary to meet the required schedule, despite the attendant high fuel consumption.

A commercial passenger-carrying airliner on the last flight of the day to a city where the aircraft will remain overnight will be on a schedule that permits slower flight, thus saving fuel, although arriving somewhat later than would otherwise be possible at another time of day.

Although highly simplistic, these examples serve to illustrate the difference between the relative worth of time and fuel under different circumstances. It is up to the operators to determine the relative value of fuel and time in a manner that will allow them to tailor their operation to yield maximum profits.

Cost index (CI)

A device used to express the relative costs of time and fuel is the cost index. Formally defined, *CI* is the ratio of time cost to fuel cost:

$$CI = \frac{\$ \ per \ hr}{\$ \ per \ lb} \tag{10-1}$$

By this definition, the higher the relative value of time, the higher the *CI*; conversely, the higher the value of fuel, the lower the *CI*.

Note that the resulting units of *CI* in Equation 10-1 are pounds per hour, which are units of fuel flow. *CI* can also be interpreted as a means of translating time resources into fuel resources in an effort to deal with common units; a high *CI* implies that a unit of time is worth relatively more units of fuel, and vice versa. By using *CI*, we are able to express both time and fuel resources in terms of pounds of fuel per hour. Having expressed both resources in the same units, we are now in a position to add them together and evaluate direct costs; therefore, without knowing the individual costs of time and fuel, we nevertheless have at our disposal their relative value and can plan the operation of the aircraft accordingly.

Time costs

The assessment of fuel costs is a relatively simple concept, and its implementation in the *CI* formulation is straightforward. The same might not be true of time cost, simply because it is made up of a number of components whose identity and relative contribution to the calculation is subject to many different accounting practices.

Selected sources of time cost that the operator must consider are:
- Crew salary and overtime
- Maintenance (only those items related to flight time)
- Lease payments (if the agreement includes payments for every hour flown)
- Insurance (also might be subject to payments based on hours flown)
- Depreciation

It is important to emphasize that the maintenance costs that will be accrued as time costs are only those expenses that can be directly attributed to the time that the aircraft spends in flight. For example, calendar-based checks must be made whether the aircraft flies or not; therefore, they do not constitute a time re-lated cost. Conversely, fuel pumps, bearings, and structural components degrade with every hour of usage, and as such their maintenance costs should be included in the time costs. Indeed, the same can be stated for any cost; only costs that change with the period of time that the aircraft is in the air should be considered.

CI variation with flight segment

Time-related costs probably do not vary with the direction of flight of the aircraft; whether the aircraft is flying from point A to point B or to point C will not have an effect on the cost of time. On the other hand, fuel cost does vary with location; therefore, it is important to have a suitable method of accounting for the variation of fuel prices for different locations such that the *CI* will reflect the cost of fuel at the location where it was purchased. Alternatively, as some operators have done, it is possible to derive a systemwide *CI* that will serve adequately for all routes, but with an associated loss in accuracy and marginally increased costs.

If the operator also practices tankering, the calculation of *CI* might become more involved, thus warranting a more detailed analysis of the tankering practices and their impact on fuel purchases.

Economy speeds

In older jet transports, *CI* can be used to calculate the direct cost of a trip flown under different conditions. Different constant cruise speeds are assumed and the direct cost (time + fuel) at the end of the trip plotted against speed. In Fig. 10-1, time cost, fuel cost, and the sum of both costs are plotted against Mach. The *optimal cruise Mach*, or *best economy Mach* (ECON Mach, or M_{OPT}), is the speed at which the sum of all costs is minimal. By varying other pertinent variables such as cruise weight, route, wind, and cruise altitude, the operator can determine the constant cruise speed that results in the minimum cost of operation.

More modern aircraft are equipped with onboard computers that can calculate the economy speed in real time based on the prevailing atmospheric conditions, aircraft weight, and the *CI* selected for the leg. In this context, the economy speed takes on three different meanings whose applicability depends on the active flight mode:

- Climb. Referring to Fig. 10-2, the economy speed is the speed that will result in the least direct cost in flying from the beginning of the enroute climb segment to a point at cruise altitude (B), beyond top of climb (TOC), where equilibrium cruise conditions have been reached.
- Descent. Similarly to the climb segment, the economy descent speed is the speed that minimizes the direct cost of flying from some point in stable cruise (C), close to the top of descent (TOD), to the point where the descent ends and the approach is initiated.
- Cruise. The economy speed for the cruise segment is the speed that minimizes the direct cost per mile.

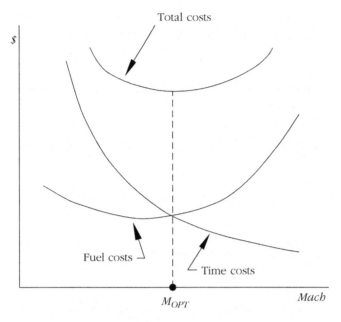

Fig. 10-1 *Time cost and fuel cost versus Mach.*

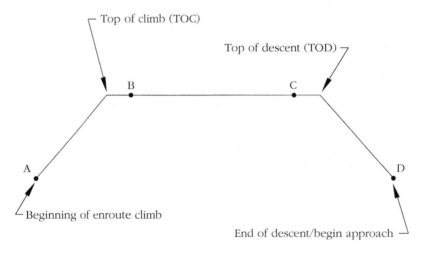

Fig. 10-2 *A simplified flight profile.*

For climb and descent, the economy speed definitions are similar to the definition used for older transports; they consider the direct costs of flying between two points and yield a constant speed.

Economy cruise mach

In cruise, the economy speed, ECON Mach, is defined in terms of an instantaneous rate: direct cost per mile. Arbitrarily, we can also elect to maximize the distance traveled per unit cost, thereby dealing with the reciprocal of the original cost-per-mile parameter; both processes are equivalent, but the distance-per-cost method will yield a presentation with which we are now familiar, namely specific range.

The direct cost in which we are interested will be expressed in pounds of fuel per hour because we will use CI to transform the cost of time into pounds per hour. We can thus write:

Total Cost = (Time Elapsed) × (CI) + Fuel Consumed, in lb per hr

or

$$Cost = T \times (CI) + F$$

We now define the economy specific range SR_E as:

$$SR_E = \frac{Rate\ of\ change\ of\ distance}{Rate\ of\ change\ of\ cost} = \frac{\Delta\ distance}{\Delta\ cost} \tag{10-2}$$

which, as stated before, is the reciprocal of the cost-per-mile function. We then seek to maximize SR_E by first rewriting Equation 10-2,

$$SR_E = \frac{TAS + V_W}{\dfrac{d}{dt}[T(CI) + F]} = \frac{TAS + V_W}{CI + f\!f} \tag{10-3}$$

If the definition of Mach number is now included, Equation 10-3 becomes

$$SR_E = \frac{a_0\sqrt{\theta}M + V_W}{CI + f\!f} \tag{10-4}$$

Substituting now the original definition of SR (Equation 9-1), Equation 10-4 transforms into

$$SR_E = \frac{a_0\sqrt{\theta}M + V_W}{CI + a_0\sqrt{\theta}\ \dfrac{M}{SR}} = \frac{SR\,(a_0\sqrt{\theta}M + V_W)}{(SR)(CI) + a_0\sqrt{\theta}M} \tag{10-5}$$

The benefit of Equation 10-5 is that it allows the representation of SR_E in terms of SR, M, V_W, and CI. Figures 10-3 and 10-4 show the variation of SR_E with M, CI, and V_W.

We can now define ECON M as the Mach number that maximizes SR_E. Alternatively, in terms of costs, ECON M is the speed that minimizes the direct cost (fuel and time costs) per mile. When we were concerned with fuel consumption only, the optimal Mach number was the speed that minimized fuel burn per mile, and we called this *maximum-range cruise*, MRC. The situation here is analogous, but the concept of specific range with which we are dealing in this section is a generalization of the original SR; we now include the value of time through the use of CI in the definition of SR_E. ECON M is shown in Figs. 10-3 and 10-4.

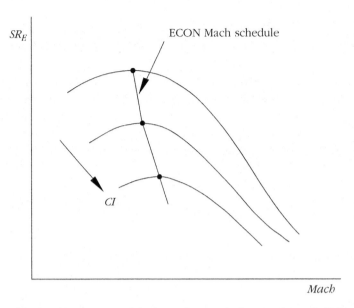

Fig. 10-3 SR_E *versus Mach and* CI *with the resulting ECON speed schedule.*

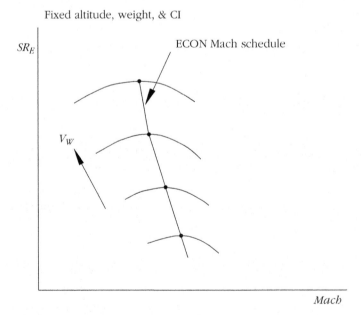

Fig. 10-4 SR_E *versus Mach and wind velocity with ECON speed schedule.*

Example 10-1

An operator of an aircraft that is not equipped with a FMS wants to know the speed at which the aircraft should be flown at FL350, FL370, and FL390, in a no-wind condition. The performance manual for the aircraft has a table of SR, as shown in the first three columns of the table below, for the weight of interest. The time-related cost of operating the aircraft is $461.50 per hour, and the average price of fuel is $0.60 per gallon.

The first step in arriving at a solution is to calculate, or look up the value of $\theta^{\frac{1}{2}}$, for later use in Equation 10-5. These values of $\theta^{\frac{1}{2}}$ are:

Flt.level	$\theta^{\frac{1}{2}}$
FL350	0.8714
FL370	0.7519
FL390	0.7514

Next we calculate the CI. Assuming that the fuel density is 6.5 lb/gal, the price per pound of fuel will be

$$\$/lb = \frac{0.60 \ \dfrac{\$}{gal}}{6.5 \ \dfrac{lb}{gal}} = 0.0923$$

The CI will then be the quotient of the times cost and the cost of a pound of fuel:

$$CI = \frac{\dfrac{\$}{hr}}{\dfrac{\$}{lb}} = \frac{461.5}{0.0923} = 5000$$

Now, using the table provided by the manufacturer, the value of SR_E can be calculated from Equation 10-5; for example, at $M = 0.66$, and FL350, $SR = 0.06672$ nam/lb, then, from Equation 10-5, repeated here for convenience

$$SR_E = \frac{SR(a_0\sqrt{\theta}M + V_W)}{(SR)(CI) + a_0\sqrt{\theta}M} \qquad (10\text{-}5)$$

$$SR_E = \frac{SR(661.5)(0.8714)(M)}{(SR)(CI) + (661.5)(0.8714)(M)}$$

$$= \frac{(0.06672)(661.5)(0.8714)(0.66)}{(0.06672)(5000) + (661.5)(0.8714)(0.66)} = 0.03555$$

The calculations for the other values are straightforward and shown in this table of specific-range values: (*Continued on page 160.*)

(Continued from page 159.)

	Specific range			Economy specific range CI = 5,000		
Mach	FL350	FL370	FL390	FL350	FL370	FL390
0.66	0.06672	0.06612	0.06463	0.03555	0.03343	0.02576
0.67	0.06737	0.06699	0.06570	0.03599	0.03388	0.02631
0.68	0.06801	0.06786	0.06680	0.03642	0.03433	0.02686
0.69	0.06863	0.06872	0.06790	0.03684	0.03479	0.02742
0.70	0.06920	0.06955	0.06898	0.03726	0.03526	0.02797
0.71	0.06972	0.07033	0.07001	0.03765	0.03574	0.02852
0.72	0.07016	0.07104	0.07097	0.03802	0.03622	0.02904
0.73	0.07050	0.07165	0.07182	0.03836	0.03669	0.02954
0.74	0.07071	0.07211	0.07252	0.03866	0.03713	0.03001
0.75	**0.07076**	0.07240	0.07301	0.03891	0.03754	0.03043
0.76	0.07061	**0.07247**	**0.07325**	0.03910	0.03789	0.03079
0.77	0.07023	0.07227	0.07319	0.03921	0.03815	0.03108
0.78	0.06957	0.07175	0.07275	**0.03922**	0.03830	0.03127
0.79	0.06860	0.07086	0.07189	0.03913	**0.03831**	**0.03135**
0.80	0.06726	0.06953	0.07052	0.03889	0.03814	0.03130
0.81	0.06550	0.06771	0.06857	0.03850	0.03772	0.03107
0.82	0.06326	0.06533	0.06596	0.03790	0.03700	0.03063
0.83	0.06050	0.06231	0.06261	0.03707	0.03589	0.02994
0.84	0.05715	0.05858	0.05843	0.03594	0.03428	0.02892
0.85	0.05314	0.05405	0.05332	0.03445	0.03201	0.02748
0.86	0.04840	0.04865	0.04718	0.03252	0.02885	0.02552

The maximum values for each column are in bold. Note how, for each altitude, the maximum value is associated with a higher Mach number when SR_E is calculated, indicating that a higher speed is optimal when time costs are included.

The accompanying plot, for FL350 only, shows the same effect pictorially. If minimum fuel consumption per nautical air mile is the only important criterion, then $M = 0.75$ is the correct speed. If the time costs that result in a CI of 5,000 are included, then the optimum speed is $M = 0.78$.

As with the original SR, SR_E can be used to define optimum altitude for any combination: of weight, CI, and wind. Indeed, many of the tools developed from the original concept of SR are equally applicable to SR_E.

As stated before, there are treatments in which the results presented here are derived following a process of cost minimization, instead of SR_E maximization. Either approach is valid, but for the professional working in the airline environment, the SR_E formulation is more useful since the performance engineering manual contains a detailed presentation of SR, thereby avoiding the need to obtain fuel flow from tables of corrected fuel flow.

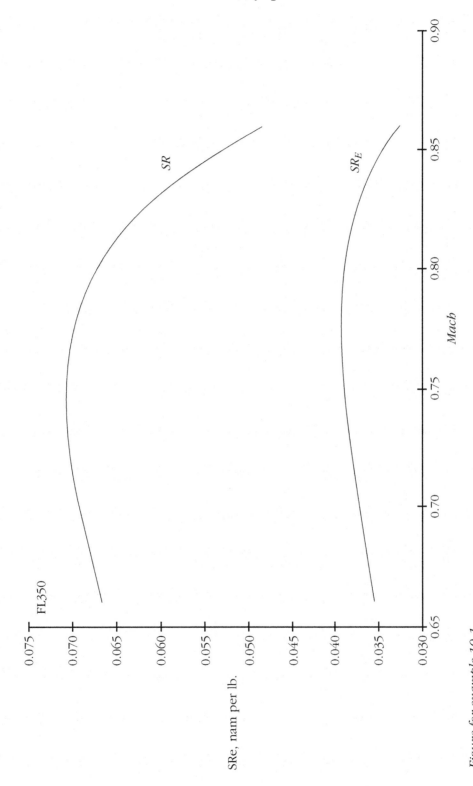

Figure for example 10-1.

Optimal trajectories

In the study of dynamic systems, we often search for the state of the system that will yield the most benefit, be it minimum cost, minimum duration, largest volume, or the like. Traditionally, engineers have solved problems of optimization by setting the first derivative of the function of interest equal to zero and then solving for the independent variable. That is the approach we have followed here.

More recently, researchers have been seeking the solution to aircraft trajectory problems in terms of the complete path, from origin to destination, in which all the state variables of the aircraft are optimally defined (weight, altitude, speed, acceleration, and energy). This is the realm of the calculus of variations and its application to optimal control theory. Solutions to these optimal trajectory problems are singularly dependent on the conditions existing at the beginning and end of the flight.

This more-sophisticated method of solution is interesting because the solution of a problem formulated as an optimal control problem is truly and uniquely optimal; for such a solution, no improvement is possible within the context of the problem definition. The solutions with which we deal in this book are not of this type; they are the next best alternative without resorting to the added complexity of a more advanced mathematical treatment. Furthermore, the more complex solutions that could be derived are not necessarily compatible with the present ATC environment; for example, a truly optimal trajectory would likely consist of a continuous climb throughout the cruise portion of the flight, accompanied by a continuously decreasing speed.

Also within the context of optimal trajectories, the concept of *CI* undergoes a change in meaning when we deal with trajectories that must end at a specified time—when the arrival time is fixed, as is the case in many of the more congested airports. In these problems, *CI* becomes a parameter that ensures that the required trajectory will also be a minimum-fuel trajectory.

These and many other ideas are the object of continuing research because they will form the basis of the ATC environment of the future, when flight management functions and flight planning functions will necessarily be integrated.

11

Descent and landing

Conceptually, the climb and descent maneuvers are similar. Furthermore, mathematically, both maneuvers have the same description; their results differ only in sign:

- A descent produces a negative rate of climb (*ROC*), or, alternatively, a positive rate of descent (*ROD*).
- A descent produces a negative climb angle, or, alternatively, a positive descent angle.

But even though the processes of climb and descent are mathematically similar, the methods of calculating the descent path are not the same for both climb and descent. The reason for this difference is that the end of the climb portion of the flight, the top of climb point (TOC), is generally free of end constraints; that is, it is not important that the TOC be at a specified distance from the origin of the flight. What is important is that the climb be executed using the speed schedule that satisfies the chosen optimality condition, be it minimum fuel, minimum cost, or minimum time. Also characteristic of the climb phase is that the progress of the aircraft along its flight path is such that, in both time and distance, the aircraft moves away from the origin, which, as we said, is the only fixed point in the climb trajectory.

Descent paths exhibit a more restricted set of circumstances—no longer do time and distance move away from the constraint. As time elapses, the distance to the destination (the constraint in descents) decreases. Indeed, it is generally unimportant where the top of descent (TOD) is located with respect to the destination point, as long as the resulting descent path ends exactly at the destination. Consequently, the planning and calculation of descent paths is more sophisticated than for climb paths. The construction of the descent path has to begin at a point such that the end of the descent path is at the desired destination; this usually calls for a process of trial and error, resulting in a number of iterations until the correct TOD is chosen.

To a computer, including the ubiquitous FMS, the problem of calculating a descent path is analogous to building a climb path because it ignores the notion that time progresses in one direction only. A computer builds the descent path from the destination backward to some point in cruise, much as it would treat a climb problem with time running forward. After arriving at the tentative cruise point where the descent calculation would be initiated, the computer compares the predicted aircraft-state variables (weight, speed, altitude, etc.) at that point with the actual values. After some error assessment, the process is repeated until the backward descent path matches the conditions of the aircraft in cruise.

In this process, the inclusion of the winds to be encountered plays a very important role in establishing an accurate descent path. Not only will a headwind have a completely different effect from what can be expected of a tailwind; the calculation is further complicated by the variation of wind speed and direction with altitude.

The effects of thrust

It is not uncommon for modern turbofan engines to produce negative net thrust during an idle descent. Of course, this thrust can also be viewed as additional drag, and is often so considered for some calculations. Whether negative or positive, the value of thrust produced by the engines is often affected by the minimum idle speed allowed in flight. Normally, the idle power setting of the engines while the aircraft is on the ground is lower than the minimum idle speed allowed in the air. Factors imposing this minimum idle speed are:

- Cabin pressurization requirements
- Power extraction to run accessories (anti-ice, bleeds, etc.)
- The ingestion of water or ice into the engine (inhibit flameout by running at higher RPM)

Indeed, the unexpected encounter of any event that requires raising the minimum idle speed can have a considerable effect on the planned descent path, making it less steep by virtue of the increased thrust. Consideration of the shallowing effect on the descent path is of paramount importance in meeting the descent restrictions imposed by either the arrival procedure in use or ATC.

The effect of drag

If thrust can make the descent path shallow, intuitively and correctly, drag has the effect of steepening the descent path and increasing the *ROD*, as will be shown in the equations presented at the end of the chapter.

Drag might be desirable during the descent. Certain situations call for increasing drag, such as increasing *ROD* to meet a restriction or a requirement to reduce speed to comply with ATC.

Drag can be increased generally by any of the following methods:

- Deploying spoilers or speed brakes
- Lowering the landing gear
- Lowering the flaps
- Using reverse thrust (uncommon technique)
- Decreasing speed (which increases angle of attack)

Any of these methods, with the possible exception of deploying the spoilers/speed brakes, is limited to speeds below which no structural damage will be incurred. Nevertheless, as with any drag-producing device, the higher the speed at which it is used, the more drag it will generate.

The effect of speed

Recall that modern turbofan engines produce more drag than thrust while in idle power, an effect that tends to be more pronounced at higher speeds. Figure 11-1 shows typical thrust available and required curves for an idle descent. The shapes of these curves can vary depending upon the installation and atmospheric conditions, but generally they will be as shown with the thrust available sloping downward as speed increases. The effect of this is to place the speed for minimum gradient, not at the lowest point of the thrust-required curve (max. L/D), but at the point where both the thrust required and thrust available curves are closest to each other.

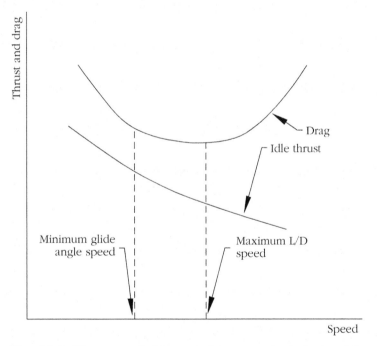

Fig. 11-1 *Thrust available and thrust required during idle descent.*

Drift-down

The drift-down maneuver consists of flying the aircraft, after an engine failure, from the original cruise altitude to some lower altitude that can be maintained

with the available thrust. Upon recognizing an engine failure, the crew will increase the power of the remaining engines to the *maximum continuous thrust*
(MCT) rating, which probably will not be enough to maintain the original cruise
altitude and speed; consequently, the aircraft will gradually slow down and
eventually begin descending. To minimize the descent angle, the crew will seek
to fly the descent at the best drift-down speed.

Figure 11-2 is a representation of the conditions found at cruise altitude at
the moment of an engine failure. At point A, the thrust available suddenly shifts
from the all-engine curve to the one-engine-inoperative (solid line), creating a
deficiency in thrust represented by B. It is possible that the drop in thrust available is not as pronounced and that the resulting thrust is represented by the
dashed line. In this case, the aircraft would lose speed until the new equilibrium
speed is reached at B'.

Recall from chapter 8 that a surplus (deficiency) of thrust results in a change
in the climb (descent) potential of the aircraft. This climb or descent potential

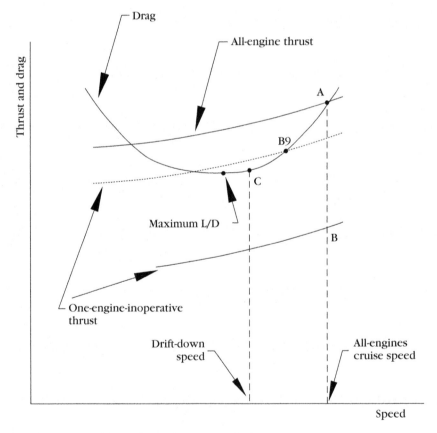

Fig. 11-2 *Cruise conditions at engine failure.*

can be expressed in terms of the *total gradient*, made up of a flight path angle and an acceleration/deceleration potential.

Again, descent is not much different from climb, with the difference being mostly arithmetic in nature—the thrust deficiency results in either a negative climb angle and/or negative acceleration. As a result, upon losing an engine at A, the crew can elect to take one of two courses of action:

- Begin descending immediately, until the thrust-available and thrust-required curves intersect again at a lower altitude.
- Remain at the same altitude, slowing down toward C where the descent must now be initiated.

The significance of point C is that it is the point on the thrust-required curve that is closest to the resulting one-engine-inoperative thrust-available curve—at C, the quantity $(D-T)$ is at its minimum value. Note that C is not necessarily at the bottom of the thrust-required curve; it is not necessarily the minimum drag point. The location of C, and therefore the definition of the drift-down speed, depends on the shapes of both the thrust-required and thrust-available curves; furthermore, these two shapes will generally change as altitude changes.

As the aircraft loses altitude, both curves will shift upward but at different rates as indicated by the arrows in Fig. 11-3. Because of this effect, the aircraft is able to reach an equilibrium altitude when the one-engine-inoperative thrust-available curve eventually reaches the thrust-required curve (Fig. 11-4). Once at the equilibrium or drift-down altitude, the crew might continue at MCT power and best-range speed (resulting in an eventual gradual climb as fuel is burned) or the power can be reduced to seek a suitable stable speed schedule, usually engine-out LRC (EOLRC).

At the equilibrium altitude, the situation is similar to that encountered in a climb with an engine inoperative. The manner in which the calculations are carried out depends on the thrust level used, whether MCT rating or thrust required.

Holding

Holding consists of flying the aircraft in a flight path that lengthens the time elapsed between the departure and arrival in such a manner that the fuel consumption rate is minimized.

Holding is usually executed while in the descent path, when, under instructions from ATC, the aircraft is held at a constant altitude. Traditionally the aircraft would be flown in a racetrack pattern or diverted to some point away from the intended flight track, usually to separate congested traffic at the destination airport and improve safety margins.

During holding, range is not important. Within the context of aircraft performance, the problem of holding is one of maximizing endurance, a topic treated in chapter 9.

Note that the maximum-endurance speeds might fall within the region of speed instability. For aircraft equipped with autothrottles, the instability is satis-

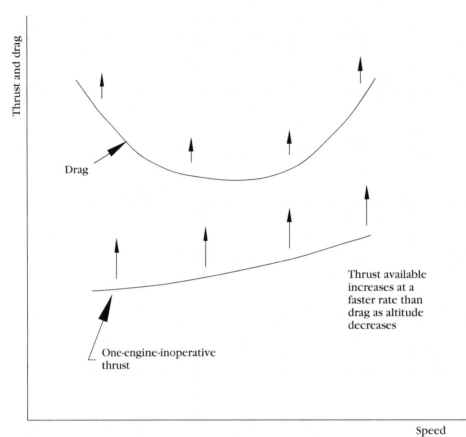

Fig. 11-3 *Thrust available and drag shift with altitude.*

factorily handled by the autothrottle computer. On the other hand, without the benefit of autothrottles, the pilot might find it difficult to maintain the desired speed, running the risk of letting the speed decrease to unsafe levels. In such cases, since the maximum-endurance speed, the maximum L/D speed, and the speed for neutral speed stability (increasing in that order) are not far apart, it would be better to fly the aircraft at a speed slightly higher than the maximum L/D speed for the current weight and altitude.

Equations of motion

Equation 8-6, which was used to describe *ROC*, can also be used to describe *ROD*:

$$ROD = -ROC = \cfrac{-V_T \cfrac{T - D}{W}}{\left[1 + \cfrac{V_T}{g} \left(\cfrac{dV_g}{dh} + \cfrac{dV_W}{dh} \right) \right]} \tag{11-1}$$

Dividing Equation 11-1 by V_T results in a good approximation to the descent angle,

$$\gamma \approx \sin\gamma = \cfrac{\cfrac{D - T}{W}}{\left[1 + \cfrac{V_T}{g} \left(\cfrac{dV_g}{dh} + \cfrac{dV_W}{dh} \right) \right]} \tag{11-2}$$

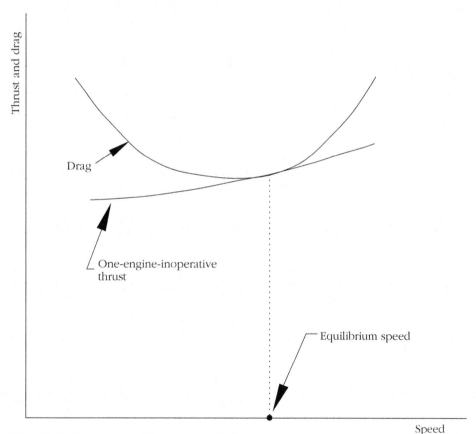

Fig. 11-4 *Descent equilibrium condition.*

As in the treatment of the climb, Equations 11-1 and 11-2 employ the concept of acceleration factor; consequently, when neither the aircraft TAS nor the wind velocity change as the aircraft progresses in its vertical path, Equations 11-1 and 11-2 are simplified because of the zero change in the ground speed and the wind speed.

Note that the wind velocity has the same effect on descents as on climbs; specifically, both the *ROD* and γ depend on the rate of change of wind velocity with altitude, not on the wind velocity itself.

The effect of weight on ROD

If we make the simplifying assumption that we have an unaccelerated descent, which does not hinder the general character of the argument to follow, Equation 11-1 reduces to

$$ROD = V_T \frac{(D-T)}{W} \tag{11-3}$$

From Equation 3-7A we know that drag can be expressed as

$$D = \frac{1}{2} C_D \rho V^2 S \tag{11-4}$$

Furthermore, from aerodynamics we know that the drag coefficient, C_D, varies as the square of the lift coefficient, C_L,

$$C_D = k C_L^2 \tag{11-5}$$

where k is a constant of proportionality with an approximate magnitude of 0.1.

If in Equation 3-6A we assume that the lift equals the weight, which is not an inordinate assumption at the shallow angles of descent, the lift coefficient can be expressed as

$$C_L = \frac{W}{\frac{1}{2}\rho V_T^2 S} \tag{11-6}$$

Substituting now Equation 11-6 into Equation 11-5 and the resulting expression for C_D substituted into Equation 11-4, the equation for drag can be written as

$$D = \frac{1}{8} \frac{kW^2}{\rho V_T^2 S} \tag{11-7}$$

Substituting now Equation 11-7 into Equation 11-3 then yields

$$ROD = V_T \frac{\left(\frac{1}{8} \frac{kW^2}{\rho V_T^2 S} - T \right)}{W} \tag{11-8}$$

Consider now the magnitudes of the parameters multiplying W^2 in the numerator of Equation 11-8:

$2k$ is in the order of $2(0.1) = 0.2$

ρ is in the order of 0.002 slugs per cubic foot

V is in the order of 510 fps (300 KTAS)
S is in the order of 900 ft^2

Therefore, the quantity

$$D = \frac{1}{8}\frac{k}{\rho V_T^2 S} \tag{11-7}$$

has a value in the order of 2.6×10^{-8}. Consequently, even though there is a square term of weight in the numerator of Equation 11-8, an increase in weight has the net effect of reducing the *ROD*. This is the result of having a multiplier of very small magnitude associated with the W^2 term and a coefficient of 1 for W in the denominator. The denominator, therefore, controls the value of the fraction, allowing a greater *ROD* for the lighter weights, which sometimes defies intuition.

The same argument can be made using calculus, differentiating Equation 11-8 with respect to W, while holding the other parameters constant.

This is a well-known effect to glider pilots (for whom thrust = 0) who carry water ballast in their aircraft to improve their cross-country performance.

A representative plot of ROD versus speed and weight is shown in Fig. 11-5.

Example 11-1

An aircraft with a wing area of 1,000 ft^2, and the following drag curve

$$C_D = 0.05 C_L^2 - 0.01 C_L + 0.02$$

is descending through 15,000 ft. at idle thrust (–500 lb.), and $M = 0.6$. What is the *ROD* if the weight is 100,000 lb., and 20,000 lb. above and below this weight?

From an atmospheric table, we can read that the density at 15,000 ft. is 0.0017 slg/ft^3, and the speed of sound is 1,057 fps. With the latter, we can calculate the aircraft TAS as

$$\text{TAS} = (0.6)(1057) = 634 \text{ fps}$$

From Equation 3-6A, we can then calculate C_L:

$$C_L = \frac{100,000}{\left(\frac{1}{2}\right)(0.0017)(634)^2(1000)} = 0.3554$$

Substituting this value of lift coefficient in the expression given above for C_D yields a value of .0212, which, through the use of Equation 3-7A yields

$$D = (0.0212)\left(\frac{1}{2}\right)(0.0017)(634)^2(1000) = 6404 \ lb.$$

From Equation 11-3, we now have

$$ROD = 634\left(\frac{5965 + 500}{100,000}\right) = 46.9 \ fps = 2817 \ fpm$$

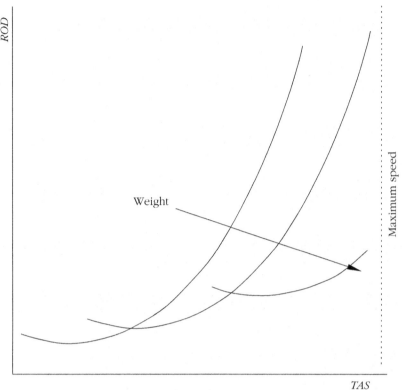

Fig. 11-5 ROD *versus TAS and weight.*

Performing the same calculations for other weights generates the results summarized in the table below:

Table 11-1 Example 11-1

W (lb)	CL	CD	D (lb)	ROD (fps)	ROD (fpm)
80,000	0.2843	0.0212	5,965	55.2	3,312
85,000	0.3021	0.0215	6,061	52.7	3,160
90,000	0.3199	0.0219	6,167	50.5	3,029
100,000	0.3554	0.0228	6,404	46.9	2,817
110,000	0.3909	0.0237	6,678	44.3	2,655
120,000	0.4265	0.0248	6,986	42.2	2,532

Landing

Just as the descent bears many similarities to the climb, the landing maneuver resembles the takeoff procedure; therefore, the calculations associated with deceleration are similar to those covered in chapter 7, with the associated changes related to deceleration and possibly negative thrust.

Climb limitations

Although seemingly incongruous with the idea of landing, there are regulations governing the climb capability of an aircraft as it prepares to land. These requirements ensure that the aircraft retains enough performance to safely abort the approach and landing process.

For the purpose of regulatory compliance, the FARs stipulate that the aircraft meet certain performance requirements while in two specific configurations: approach climb and landing climb. These configurations are defined in Table 11-1.

**Table 11-1 Landing configurations
with speed and gradient requirements**

	Approach climb	Landing climb
Gear	Up	Down
Flaps	Approach setting.	Landing setting.
Power	One engine inop; takeoff thrust on remaining engines.	Takeoff thrust on all engines.
Speed	1.5 V_{stall}	1.3 V_{stall}
Gradients (%):		
Two-engines	2.1	3.2
Three-engines	2.4	3.2
Four-engines	2.7	3.2

These climb requirements translate into weight limitations that are normally tabulated for a particular flap setting as a function of ambient temperature and airport elevation.

In contrast to the takeoff requirements, there is no altitude specification imposed on either the approach-climb or the landing-climb configurations; neither is there a requirement to overfly, by a given margin, any obstacles that lie under the flight path. Furthermore, there is no statement addressing the condition resulting after one engine becomes inoperative while in the landing-climb configuration, the implicit assumption being that the aircraft is committed to a landing in this circumstance.

Individual operators usually assume the responsibility of providing additional training in this regime of flight. They also provide specific information on obstacle avoidance when the buildings and terrain surrounding the airfield merit the additional attention. To provide this information, the operator calculates the climb capability of the aircraft using the same equations utilized in dealing with the takeoff problem but considering the additional drag produced by the landing gear and the flaps in a landing configuration.

Required landing distance

The demonstrated landing distance is the distance required by the aircraft to transition from the landing configuration, 50 feet above the runway, to a complete stop on the runway. Furthermore, the FARs require that an aircraft have available for landing a distance equal to 1.67 times the demonstrated landing distance for the weight and speed under consideration.

For wet runways, the landing distance requirement must be increased by 15 percent, unless the manufacturer can demonstrate better performance during certification. These concepts are illustrated in Fig. 11-6.

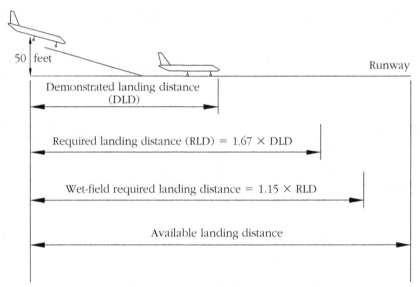

Fig. 11-6 *Landing distance definitions.*

Maximum quick turnaround weight

The capability of the brakes to stop an aircraft on the runway depends on their capacity to dissipate heat. Naturally, to some extent, the ambient temperature is a determining factor in the time required to dissipate any amount of heat. But more importantly, the operator will be interested in the amount of heat generated by a landing, especially when a takeoff will be attempted shortly after landing. The heat generated when stopping an aircraft, in turn, is strongly influenced by the ambient temperature, the airport altitude, and the aircraft weight. High-altitude landings by a heavy aircraft with a shallow flap setting in relatively warm air temperatures will require fast landing speeds—this will translate into a high kinetic energy that needs to be absorbed by the brakes.

For any combination of landing weight, flap setting, airport altitude, and ambient temperature, the manufacturer tabulates the length of time over which the brakes should not be used before another takeoff is attempted. The importance of this waiting period is underscored by the recognition that the takeoff calculations are predicated on maximum-capability braking; this level of stopping power is not available unless the brake mechanism has been allowed to cool per the prescribed schedule.

Overweight landings

There are many reasons why an aircraft can be landed at a weight heavier than normally allowed. Most of these reasons pertain to unforeseen events that force the crew to land without the benefit of the aircraft weight by dumping fuel.

A consideration in such situations is the braking ability of the aircraft at the heavier weight. If the length of the proposed runway does not provide ample distance to execute the landing without dangerously approaching the threshold of a brake fire, the landing should be attempted at another runway. Additionally, the heavier-weight landing will require faster speeds, which will subject the flaps to additional aerodynamic loads, perhaps even beyond the flap placard speeds.

Finally, the rate of descent of a heavier-than-normal landing will result in sink rates that produce additional impact loads on the structure.

All of these effects call for a mandatory inspection of the aircraft after landing at a heavy weight.

12

Weight and balance

All aircraft have inherent limitations of weight that apply at different stages of the operation, and equally there are limitations on the location of the center of gravity (CG) of the aircraft. Both of these limitations are presented in the aircraft flight manual (AFM) in a graph similar to the one shown in Fig. 12-1, known as the weight-CG envelope, or certified envelope.

Shown in Fig. 12-1 are not only the limiting conditions of weight/CG, but also the maximum weights that the aircraft can have under specific conditions:

- The *maximum takeoff weight* (MTOW) is the maximum weight allowable for any takeoff, regardless of the performance capabilities of the aircraft.
- The *maximum taxi weight* (MTW), also known as the maximum ramp weight, is the maximum weight that the aircraft can have under any circumstance.
- The *maximum landing weight* (MLW) is the maximum weight at which the aircraft is allowed to land.
- The *maximum zero-fuel weight* (MZW) is the maximum weight of the aircraft without useable fuel.

Additionally, the forward and aft edges of the envelope limit the location of the aircraft CG. Often, as shown, there will be separate limits for different phases of flight, sometimes in both edges of the envelope, not only in the forward edge as in the case shown.

These boundaries are established by the manufacturer and reflect the inherent limitations of an aircraft to operate in conditions that would render it unstable or result in structural damage. It is incumbent upon operators, therefore, to conduct their operations in such a way that the envelope is not violated. Normally this is accomplished by designing a loading system for the aircraft that is tailored to the operator's procedures. The loading system will consist of a curtailed envelope plus the tables that the operator will use in calculating the effect of passengers, fuel, and cargo on aircraft CG.

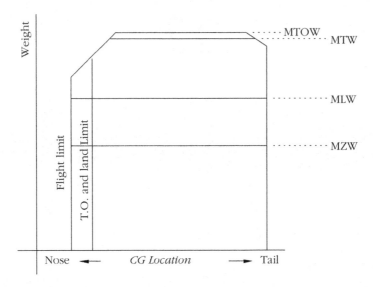

Fig. 12-1 *A generic example of a certified envelope.*

Operational allowances
and the curtailed envelope

In the operation of a jet transport, there are many sources of uncertainty as to the location of the aircraft CG that must be accommodated within the certified CG envelope. These uncertainties arise from the methods used to load passengers and cargo, as well as the movements of payload items and fuel while in flight. Consider, for example, the following items:

- Even though all passengers might be allocated certain seats, it is possible that some of them will not be seated in their assigned seats at the time of takeoff, and knowledge of their exact location will remain unknown at the time of calculating the takeoff parameters.
- While in flight, carts weighing as much as 200 pounds carrying beverages and meals will be rolled up and down the cabin, producing some change in the location of the CG.
- As fuel burns in a swept-wing aircraft, the location of the aircraft CG necessarily changes because most if not all the fuel is stored in the wings.

These are only some of the sources of CG variation that an aircraft will experience in normal operation. Other sources might be the movement of flaps and landing gear, the location of cargo, and fuel dumping. Merely loading the aircraft in such a way that the resulting CG falls within the certified envelope is not enough to ensure that a violation of the envelope will not occur later.

A salient effect is that of fuel as it is burned during flight. As a consequence of the shape of the wings and the fuel tanks within them, the depletion of fuel has the effect of moving the CG of the aircraft first forward, up to a point (C in Fig. 12-2), and eventually backward. If the aircraft were loaded such that the resulting CG at takeoff were close enough to the forward edge of the certified envelope (point B), the forward movement of the CG due to fuel consumption would eventually cause the forward edge of the envelope to be violated (the dotted line in Fig. 12-2).

To preclude the occurrence of such events, the operator will make allowances for these contingencies. These allowances take the form of curtailments to the certified envelope. In Fig. 12-3, the forward edge of the certified envelope has been moved back enough to accommodate the uncertainties associated with the location of the CG while operating the aircraft. In the example of the fuel consumption, the amount of curtailment might be equal to the CG

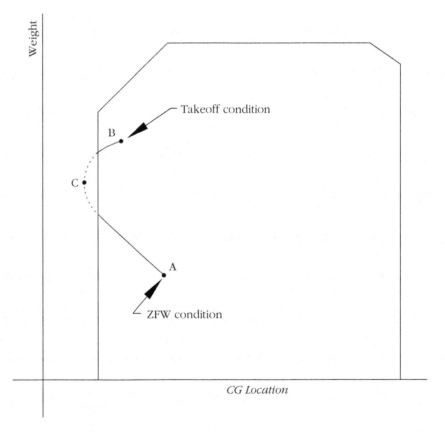

Fig. 12-2 *Movement of aircraft CG as fuel is consumed in a swept-wing transport aircraft.*

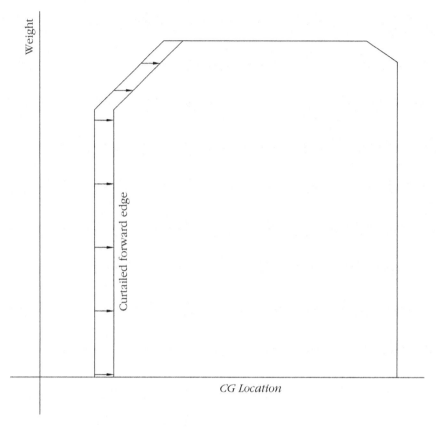

Fig. 12-3 *Curtailment of the forward edge of the certified envelope.*

shift between points B and C. In other words, if the operator limited the loading of the aircraft—not by the original certified envelope, but by a new curtailed envelope (Fig. 12-3)—the worst possible outcome would be that the CG might end up at the edge of the certified envelope sometime in the course of the flight.

Fuel is not the only allowance that is considered. The meal carts, movement of passengers, and many other circumstances must also be considered, each contributing a further curtailment of the envelope. And not only is the forward edge of the envelope subject to curtailment, the aft edge is equally treated to account for those effects that might move the CG aft. The resulting envelope is called the *curtailed* or *operational envelope.*

The curtailed envelope, much narrower than the certified envelope, is the guide used for loading the aircraft in normal operation. Its purpose is to allow the operator to load the aircraft in such a way that any foreseeable event that might normally take place between taxi-out at the origin airport and landing at the destination will not cause a violation of the certified envelope, even if there are CG excursions outside the curtailed envelope. The key words here are "foreseeable event." It is the responsibility of the operator to establish and quantify those events that might result in CG shifts, but because different operators will use their aircraft differently, the shape and size of the curtailed envelope will depend on the operation, and, naturally, on the aircraft.

Each allowance and its associated curtailment will of course narrow the resulting curtailed envelope further, and, although the resulting envelope will be safe, it will also reduce the flexibility with which the aircraft can be loaded. For example, the aircraft operating under the narrow curtailed envelope will be less accommodating to last-minute cargo and passengers that could otherwise be loaded. A judicious establishment of the allowances and their curtailments is therefore necessary for the conduct of a profitable operation.

Example 12-1

A passenger aircraft has a certified envelope that, at a weight of 100,000 lb., restricts the CG to be within the forward limit of 620 inches and the aft limit of 650 inches; the distances are from the reference datum defined for the aircraft. Calculate the curtailment necessary to allow for a cabin of 100 seats in which the first-row passengers would be seated at an arm of 300 inches and the last-row passengers at 970 inches; assume that the operator expects 10 percent of the passenger load to be seated in seats other than those assigned to them. Each passenger and carry-on baggage is assumed to weigh a total of 180 lb.

Solution: First we locate the centroid of the seats, which we assume to be uniformly distributed in the cabin; therefore, the middle of the set of seats is at a point midway between 300 inches and 970 inches

$$\overline{x}_{cabin} = \frac{300 + 970}{2} = 635"$$

The middle of the cabin is at 635 inches, which means that 10 percent of the passengers (10 PAX) could be seated forward of the 635-inch line, and another 10 PAX could be seated behind the same line. Lacking more detailed information, we must assume that the 10 PAX relocate themselves to the centroid of the half of the cabin they have selected; in other words, they move to the middle of the forward or aft half of the entire cabin, which requires that we calculate those centroids as well:

For the forward half, the centroid is located at

$$\overline{x}_{fwd} = \frac{300 + 635}{2} = 467.5"$$

and for the aft half,

$$\overline{x}_{aft} = \frac{970 + 635}{2} = 802.5"$$

When 10 PAX move from the middle of the cabin to their new location at the middle of the forward half, the moment associated with that shift is

$$M_{fwd} = (10)(180)(467.5 - 635) = -301,500 \; inch\text{-}lb$$

where a negative sign indicates that the moment is associated with a nose-down tendency. A similar calculation would reveal that the moment associated with 10 PAX moving to the aft half of the cabin is also 301,500 in-lb., but with a positive sign.

There is a need, then, to curtail the forward edge of the envelope by an amount that will accommodate a moment of 301,500 in-lb., or

$$\overline{x} \pm \Delta\overline{x} = \frac{\Sigma W_i x_i}{\Sigma W_i} \pm \frac{\Delta M}{\Sigma W_i} = \frac{\Sigma W_i x_i \pm \Delta M}{\Sigma W_i}$$

Whether the plus or the minus sign is used depends on which edge of the envelope is to be curtailed. For the forward edge, which needs to be moved aft, the plus sign is used. For the aft edge, which needs to be moved forward, the minus sign is used.

We now consider the aircraft—loaded with passengers and weighing 100,000 lb.—at the forward edge of the envelope (620 inches) and calculate the required curtailment,

$$(100,000)(620) = 62,000,000 \text{ in-lb}$$
$$+ \; 301,500$$
$$= 62,301,500$$

If the resulting moment is then divided by the weight of the aircraft, we obtain the new location for the forward edge of the envelope,

$$\frac{62,301,500}{100,000} = 623.02"$$

Suppose that while observing the new operational envelope, the operator loads the aircraft conservatively, and the CG falls at 625 inches, almost a full 2 inches inside the curtailed envelope. Suppose further that at the last minute the loadmaster wants to load a 1,000-lb. box (for which the sender is paying handsomely as a last-minute load) in the forward cargo hold (centroid at 400 inches). The moment associated with this load is

$$(1,000)(400) = 400,000 \text{ in-lb}$$

and the new location of the aircraft CG is

$$\frac{(100,000)(625) + (1,000)(400)}{101,000} = 622.77"$$

which is forward of the forward edge of the curtailed envelope. The package could not be loaded without violating the curtailed envelope that the carrier has calculated for its operation. Note, though, that the certified envelope's forward edge is at 620 inches and, therefore, would not be violated by loading the package.

Maybe the curtailed envelope can be expanded. Suppose that the operator elects to split the cabin into two halves for the purpose of counting passengers. Now we can consider not one, but two cabins, each containing 50 seats. The forward cabin (cabin 1) has a centroid at 467.5 inches, and the aft cabin (cabin

2) has a centroid at 802.5 inches, which we already calculated. Each cabin will now be subjected to the same exercise of shifting 10 percent of its passengers forward and aft within it and establishing the resulting moments to determine the required curtailment. Following the same procedure as before, the forward and aft cabins are divided into halves to calculate the effect of moving 10 percent of the passengers. The moment associated with moving forward 10 percent of the passengers in the forward cabin is

$$M_{1\,fwd} = (5)(180)(383.75 - 467.5) = -75,375\ inch.lb$$

Similarly, because of the symmetry of our problem, all the other moments are also 75,375 in-lb., varying only in sign, depending on the direction of movement, forward or aft:

- $M_{1\,aft}$ = 75,375 in-lb.
- $M_{2\,fwd}$ = –75,375 in-lb.
- $M_{2\,aft}$ = 75,375 in-lb.

We now calculate the required curtailment under the new arrangement:

$$(100,000)(620) = 62,000,000\ \text{in-lb.}$$
$$+75,375\ \text{from}\ M_{1\,fwd}$$
$$+75,375\ \text{from}\ M_{2\,fwd}$$
$$= 62,150,750$$

The new location for the forward edge of the curtailed envelope is then

$$\frac{62,150,750}{100,000} = 621.50"$$

The 1,000-lb. package that arrived at the last minute can now be loaded because the forward edge of the new curtailed envelope allows the resulting CG to be located at 622.7 inches, which is behind the new forward edge.

The process of curtailing the certified envelope is complex and depends on the aircraft whose envelope is being curtailed in concert with the operation to which it will be subjected. Example 12-1 illustrates the necessary process, but is deceiving in its simplicity. Notably complicated is the curtailment necessary to guard against envelope violations due to fuel consumption. The curve depicting the travel of aircraft CG as fuel is burned is often complex and will depend on the sequence in which the tanks are depleted and even on the possibility of fuel dumping.

Example 12-2

The forward edge of the certified envelope for an aircraft lies at 634.50 inches from the datum for weights up to 110,000 lb. From there, it varies linearly to 115,000 lb., where the envelope is at 635.00 inches. The fuel contained in the wing tanks varies in location according to the first two columns of the table below. Calculate the variation in aircraft CG as the fuel is loaded from a ZFW of 95,000 lb. Furthermore, determine the most critical fuel quantity as it pertains to violation of the certified envelope.

Table for example 12-2
ZFW = 95,000
CG@ 636.00

Fuel weight, lb.	Fuel CG, inches	CG, inches	Aircraft weight, lb.	
0		636.00	95,000	ZFW
500	605.74	635.84	95,500	
1,000	607.89	635.71	96,000	
1,500	609.85	635.59	96,500	
2,000	611.65	635.50	97,000	
2,500	613.28	635.42	97,500	
3,000	614.77	635.35	98,000	
3,500	616.11	635.29	98,500	
4,000	617.32	635.25	99,000	
4,500	618.42	635.20	99,500	
5,000	619.40	635.17	100,000	
5,500	620.29	635.14	100,500	
6,000	621.09	635.11	101,000	
6,500	621.81	635.09	101,500	
7,000	622.46	635.07	102,000	
7,500	623.06	635.05	102,500	
8,000	623.61	635.04	103,000	
8,500	624.12	635.02	103,500	
9,000	624.61	635.01	104,000	
9,500	625.09	635.01	104,500	
10,000	625.56	635.01	105,000	
10,500	626.03	635.01	105,500	
11,000	626.53	635.02	106,000	
11,500	627.05	635.03	106,500	
12,000	627.60	635.06	107,000	
12,500	628.21	635.09	107,500	
13,000	628.87	635.14	108,000	
13,500	629.60	635.20	108,500	
14,000	630.41	635.28	109,000	
14,500	631.31	635.38	109,500	
15,000	632.31	635.50	110,000	
15,500	633.42	635.64	110,500	
16,000	634.65	635.81	111,000	
16,500	636.01	636.00	111,500	
17,000	637.52	636.23	112,000	Full fuel

The solution to the problem is easily arrived at by recognizing that the aircraft-plus-fuel CG can be calculated with the expression

$$X_{aircraft} = \frac{ZFW\,(X_{@ZFW}) + W_{fuel}X_{fuel}}{ZFW + W_{fuel}}$$

The two columns to the right in the table for this example show just such a calculation between ZFW and full tanks. This is depicted pictorially in the Example 12-2 figure, where the forward edge of the envelope has been included.

From the figure, it is clear that the aircraft CG is closest to the forward edge of the envelope at a weight of about 105,000 lb., when the fuel load is about 10,000 lb.

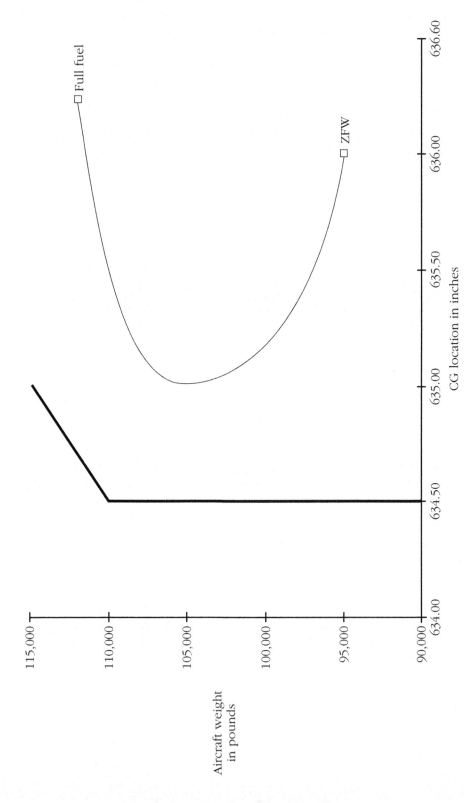

Figure for example 12-2.

13

Payload-range
curves

Often in the study of aircraft performance the operator needs to assess the capability of the aircraft in terms of both the payload that the aircraft can carry and how far the aircraft can fly with this payload. Such needs arise perhaps when the airline considers new equipment and peruses the documentation provided by the manufacturer or when existing equipment in the fleet is considered for missions other than those to which they are committed. This chapter will visit this topic and study the presentation of payload and range information.

Payload-fuel curves depict limited resources

The payload-fuel curve is a variation of the limited-resources curve often seen in presentations dealing with economics—naturally so since the curve depicts the relationship between two scarce resources, namely payload and fuel. In other words, the aircraft has a limited lift capability, whether this be established by takeoff performance limitations or the certified MTOW. This lift capability can be used to lift some combination of payload and fuel, the aggregate of which is limited. The curve depicts a constant, which is the sum of the weight of fuel and the weight of the payload,

$$fuel + (a)(payload) = constant \qquad (13\text{-}1A)$$

or

$$F + aP = C \qquad (13\text{-}1B)$$

where a is a coefficient that determines the slope of the line and depends on the units used for payload and fuel, and *constant* is the total lift capability of the aircraft under the conditions of interest up to a maximum of MTOW. The equation can be plotted as shown in Fig. 13-1.

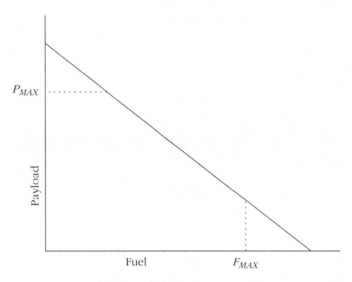

Fig. 13-1 *Payload and fuel are two scarce resources whose sum is a constant.*

The limitations imposed by the internal distribution of space within the structure of the aircraft actually limit further the amounts of payload and fuel that can be carried. These limitations are represented by the dotted lines at P_{max} and F_{max} in Fig. 13-1. That an aircraft can lift, besides the OEW, 250,000 lb. at takeoff does not mean that it can carry no payload and a fuel load of 250,000 lb.; there probably isn't enough room in the tanks to carry that much fuel. Alternatively, carrying just enough fuel to take off and execute an immediate landing doesn't allow the payload to reach 250,000 lb., again because there probably isn't enough room within the fuselage to carry that much weight without violating the load distribution constraints of the structure.

Development of payload-range curves

For practical use, the amount of fuel carried is not valuable in assessing aircraft performance. What is desired is knowledge about the range of the aircraft. We seek then to effect a transformation of the horizontal axis in Fig. 13-1 so that it can be read in terms of range.

We can arbitrarily replace the fuel scale with a distance scale, knowing that both variables vary monotonically with time, and that, therefore, there is a one-to-one correspondence between them. Also, as a result of this change, the resource curve will undergo a transformation that needs to be investigated.

Equation 13-1 can be rewritten as

$$aP = C - F$$

and then differentiated to produce the following expression,

$$a\frac{dP}{dX} = -\frac{dF}{dX} = -\frac{1}{SR}$$

which, as we can see from the above expression, conveniently develops into an expression for *SR*, a tool with which we are now familiar—specific range.

SR, as used in this context, is not the instantaneous value at origin or destination; instead, it is a value representative of the efficiency of the entire flight at the speed and altitude for which the payload-range curve was specified.

Since *SR* decreases with weight and weight decreases with range, we can expect that—for the higher payloads and shorter ranges when the weight of the aircraft will not diminish much during the flight—the *SR* will be relatively low, and, therefore, the slope *dP/dX* will be steeper (lower *SR*). This is so because a short range implies a relatively low fuel load, and therefore less of a change due to fuel burn (Fig. 13-2). For a lower-payload, longer-range flight, when more weight will be lost due to fuel burn, *SR* will increase toward the end of the flight, thereby decreasing the slope *dP/dX* (higher *SR*).

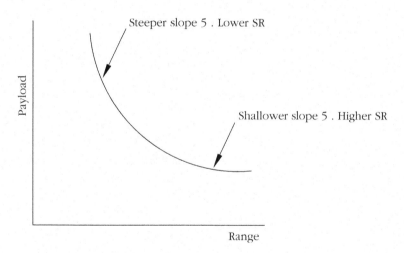

Fig. 13-2 *The payload-range curve comes from the payload-fuel curve through a transformation that uses* SR.

In an attempt to continue the exchange of payload for fuel along the maximum weight curve, we can reduce the payload further and, simultaneously, increase the fuel load by the same amount. Eventually the maximum fuel line is reached (F_{max} in Fig. 13-1, and A in Fig. 13-3), and any further reduction in payload cannot be accommodated with increases in the fuel load.

In essence, we have now a new total resources constant C_2, in Equation 13-1B. No longer can we extend the range of the aircraft by sacrificing payload to gain in fuel. Beyond this point, any further reductions in payload merely reduce the

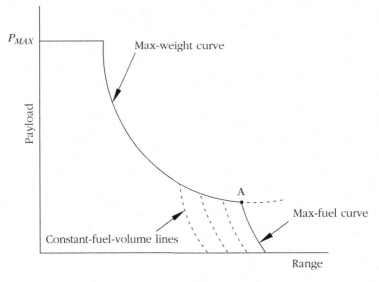

Fig. 13-3 *The maximum-fuel limitation takes the form of a curve in the payload-range plot.*

total weight of the aircraft, with the attendant increase in *SR*. In the payload-range presentation, note that what was a straight line in the payload-fuel line at the maximum fuel point, F_{max}, is now a curved line exhibiting a shallower slope as the total weight of the aircraft decreases. This is an expected result because the lower weight aircraft will have better range efficiency, or higher *SR*, and therefore lower slope of its payload-range curve.

It must be noted, nevertheless, that in some presentations the curvature in the lines is not evident. Such a curvature is a characteristic of the aircraft under consideration. For some aircraft, the lines in the payload-range curve are almost straight, and such is the case for the hypothetical aircraft that we will use in the example problems.

Example 13-1

If an aircraft (with the payload-range plot shown) has a performance-limited takeoff weight that equals the MTOW, what is the maximum payload that can be carried to a distance of 1,600 nautical miles?

The 1,600 nautical-mile range is available precisely at the point where the maximum payload line meets the MTOW line and also at the same point where the 180,000-lb. fuel-load line is located; therefore, the payload carried is maximized, and the fuel required to traverse the distance is 180,000 lb.

Now, it must be kept in mind that no flight can be undertaken in the knowledge that no fuel will be left at arrival, which is the case in the solution given so far. More realistically and by regulation, some reserve fuel must be in the tanks at the destination. If we require that 5,000 lb. of fuel be in the tanks at arrival,

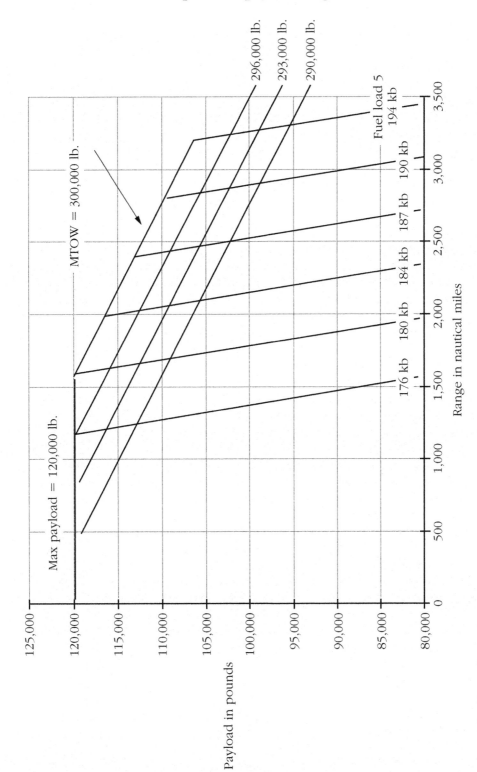

Figure for example 13-1.

the picture changes somewhat because, in essence, we are imposing the additional requirement that the payload account include this reserve fuel. The actual payload is therefore reduced by the amount of reserve fuel.

Example 13-2

For the same aircraft, if the payload that needs to be delivered at the destination is 105,000 lb., with a reserve fuel load of 8,000 lb., what is the range expected?

First we augment the payload by the amount of reserve fuel required, so to the 105,000 lb. we add the reserve fuel to arrive at a total of 113,000 lb. At MTOW, a value of 113,000 lb. in the payload scale corresponds to approximately 2,400 nautical miles.

Example 13-3

Our example aircraft is to be used in delivering emergency supplies to any of a number of distant airports in the same country, the nearest of which is at a range of 2,700 nautical miles, and the farthest is 3,500 nautical miles away. It is desired to maximize the payload, while retaining a 5,000-lb. fuel reserve, and reach as far as possible. What are the options available?

At a range of 2,700 nautical miles and an MTOW of 300,000, the payload scale reads 111,000 lb. Subtracting the 5,000-lb. reserve leaves us with a payload of 106,000 lb. The fuel required would be about 188,000 lb.

Extending the range to 3,250 nautical miles, the payload scale reads 106,500, which yields an actual payload of 101,500 lb. The fuel required would be the maximum capacity, or 194,000 lb. But, note that since we have reached the maximum fuel capacity, the reserve fuel that we have to carry must be included in the quantity carried by the tanks; therefore, the actual maximum fuel capacity, if we want to use 5,000 lb. for reserve, is 194,000 – 5,000, or 189,000 lb. At a maximum usable fuel capacity of 189,000 lb., following along the MTOW line, we read that the available payload is slightly over 111,000 lb. We have not improved much over the last iteration.

Extending the range farther will not allow us to follow the previous procedure of following the MTOW line. Any further increases in range will have to come solely from reducing payload. Furthermore, we cannot even use the 194,000-lb. line of maximum fuel capacity because we need to include the 5,000-lb. reserve, reducing the actual maximum tank capacity to 189,000 lb.

The requirement of reserve fuel produces a double penalty. First, the payload has to be reduced by the amount of the reserve fuel. Second, when the full capacity of the tanks is needed, we need to curtail that capacity by the amount of reserve fuel carried.

Note also that at extreme ranges, when the full capacity of the tanks is required, any change in payload produces only a small change in range, an effect that is due to the slope of the fuel capacity lines. While operating along the MTOW line (or any takeoff-weight limiting line), a change in payload resulted in a much greater increase in range, again because of the slope of the MTOW line, which, in turn, is an effect of exchanging payload for fuel along that line.

Equipment evaluation

The payload-range curve is a valuable tool in evaluating new aircraft for the fleet. Whether the aircraft comes directly from the assembly line, or is procured from another operator, manufacturers will provide the prospective operator with a variety of data to be used in the evaluation of the new equipment in terms of its suitability to the operator's fleet. The payload-range curve is one of the most common presentations. It depicts in a concise format some very specific capabilities of the aircraft, where the word *specific* needs to be emphasized.

The operator needs to be aware that the payload-range curve pertains to a very specific climb-cruise-descent regime, both in cruise speed (usually LRC) and cruise altitude, as well as some generic takeoff, climb, descent, and landing conditions. Whether this set of conditions is pertinent to the manner in which the aircraft will be operated is to be decided by the fleet administrator. Furthermore, and of paramount importance, the payload-range curve does not address any issues arising from any but the most generic performance takeoff limitations. It is not uncommon for an eager salesman to present an aircraft as being capable of carrying a substantial payload in a long-range route, without paying attention to the takeoff conditions prevailing at the operator's base.

For instance, both payload and fuel (and therefore range) might have to be curtailed substantially if the operator intends to base the aircraft at an airport where density altitude is relatively high (high deserts in the summer) and the runway is not long enough to accommodate the speeds that would be necessary to negotiate a successful takeoff if an engine fails. Alternatively, the aircraft might not have the braking capability to allow it to stop in the distance allotted from the necessarily high V_1.

Another common constraint is noise. Although all new aircraft are required to meet strict noise limitations, some aircraft have exhibited better characteristics than others under varying conditions; in turn, these characteristics translate directly into takeoff weight.

Yet another takeoff limitation is made up by the set of obstacles that an operator might use in calculating takeoff for the fleet. The generic takeoff used by the sales personnel usually takes into account no obstacles and no turns before the aerodynamic cleanup is completed (retracting the landing gear and any airfoil components that improve lift during takeoff).

The alert operator will discern these nuances and realize that the payload-range curve that needs to be studied is not necessarily the one depicting MTOW as the limiting weight but some lower value that corresponds to the actual performance-related takeoff limitations.

Abbreviations

α	angle of attack
β	angle of sideslip (yaw)
γ	angle, specific heat ratio
δ	pressure ratio
δ_t	total pressure ratio
Δ	change
θ	angle
θ	temperature ratio
θ_t	total temperature ratio
μ	viscosity, friction coefficient
ρ	density
σ	density ratio
ϕ	angle
a	acceleration
b	reference length
AC	advisory circular
AF	acceleration factor
AFM	airplane flight manual
AGD	accelerate-go distance
AR	aspect ratio
ATC	air traffic control
amb	ambient condition
ASD	accelerate-stop distance
$BEOC$	best engine-out climb speed
BFC	balanced field condition
CAS	calibrated airspeed
C_D	drag coefficient
CDL	configuration deviation list
CG	center of gravity
CI	cost index
C_L	lift coefficient
$C_{L\alpha}$	rate of change of C_L with angle of attack
C_m	moment coefficient

D	drag
EAS	equivalent airspeed
ECON	economy speed cruise mode
EOLRC	engine-out LRC
EPR	engine pressure ratio
ETP	equal time point
FARs	Federal Aviation Regulations
ff	fuel flow
FL	flight level
FMC	flight management computer
FMS	flight management system
FN (F_n)	net thrust
FNC (F_{nc})	corrected net thrust
fps	feet per second
ft	feet
F_x	force in the x-direction; similarly used for y- and z-
g	gravitational acceleration, 32.2 fps/second
GA	general aviation
GS	ground speed
HCM	horizontal clearance method
HLD	high-lift devices
b_p	pressure altitude
IAS	indicated airspeed
ICAO	International Civil Aeronautics Organization
IR	integrated range
KCAS	knots calibrated airspeed
KEAS	knots equivalent airspeed
KIAS	knots indicated airspeed
KTAS	knots true airspeed
l	length
L	lift
LE	leading edge
LOC	localizer
LRC	long-range cruise
LTOA	long trip optimum altitude
m	mass
MAC	mean aerodynamic chord
MCL	maximum climb thrust rating
MCR	maximum cruise thrust rating
MCT	maximum continuous thrust rating
MEL	minimum equipment list
MLH	maximum level-off height
MRC	maximum range cruise

MSL (msl)	mean sea level
MLW	maximum landing weight
MTOW	maximum takeoff weight
MTW (MTXW)	maximum taxi weight
MZFW	maximum zero-fuel weight
n	ratio of force to weight or load factor
N_1	fan rotational speed
N_2	compressor rotational speed
N_3	compressor rotational speed
NAM (nam)	nautical air miles
NM (nm)	nautical miles
OAA	obstacle accountability area
PAX (pax)	passengers
PNR	point of no return
p_s	static pressure
p_t	total pressure
q	dynamic pressure
r	radius
R	radius; universal gas constant = 53.3
ROC	rate of climb
ROD	rate of descent
RPM	revolutions per minute
RTO	rejected takeoff
s	length, distance
S	reference surface area
SID	standard instrument departure
SL	sea level
SR	specific range
STAR	standard terminal arrival
STOA	short trip optimum altitude
STP	standard temperature and pressure
t	time
T	thrust, temperature
TAS	true airspeed
TE	trailing edge
TIT	turbine inlet temperature
TO	takeoff
TOC	top of climb
TOD	top of descent; takeoff distance
TOR	takeoff run
TOW	takeoff weight
T_s	static temperature
TSFC	thrust specific fuel consumption

TSP	thrust setting parameter
T_t	total temperature
V	velocity, speed
VCM	vertical clearance method
V_1	speed beyond which takeoff cannot be rejected
V_2	takeoff safety speed
V_c	calibrated speed
V_e	equivalent speed
V_{ef}	engine failure speed
V_g	ground speed
V_i	indicated speed
V_{LOF}	lift-off speed
V_{mbe}	maximum brake energy speed
V_{mca}	minimum control in the airspeed
V_{mcg}	minimum control on the ground speed
V_{mu}	minimum unstick speed
V_R	rotation speed
V_s	stall speed
V_{sn}	stall speed at a load factor n
V_t	true airspeed
V_{te}	maximum rotational speed for the tires
V_W	wind velocity
W	weight
W_f	fuel flow
W_{fc}	corrected fuel flow
WSF	wind shift function
$\Rightarrow$	implies
$\approx$	approximately
$^\circ$C	degrees Celsius
$^\circ$F	degrees Fahrenheit
$^\circ$K	degrees Kelvin
$^\circ$R	degrees Rankine

References

Jet Transport Performance Methods, by operations engineering staff of the Boeing Company; Boeing, Seattle, May 1989.

Thermodynamics, by V.M. Faires; Macmillan Company, London, 1970;

Airplane Aerodynamics, by S.S. Sherby, T.F. Connolly; Pitman Publishing Corp., New York, 1967.

Jet Airplane Performance, by Lufthansa Consulting GmbH, Lufthansa German Airlines, Cologne, 1988.

Index

A

above ground level (AGL), 5
absolute ceiling, 109
accelerate-stop distance (ASD), 66
acceleration factor, 101
    corrections for ROC, 102-106
acceleration gradient, 113-115
aerodynamics, 13-20
    compressible/incompressible flow, 13-14
    parameters, 17-18
airfoils, 21-25
    distribution of pressure over, 18-20
    properties, 21-22
altitude, 5-6
    cross-over, 105
    effects on cruise, 139-142
    effects on ROC, 107-109
    geopotential, 6
    long trip optimum, 144
    short trip optimum, 143-144
ambient density, 4
ambient pressure, 2-4
ambient temperature, 2
angle of attack, 21, 23
angle of climb, 109-115
atmosphere, 1-6

B

balanced field condition (BFC), 68
bank-angle
    influence on climb gradient, 49-53
    influence on G-loads, 40-41
    influence on turn radius, 47-49
best economy Mach, 155
boundary layer, 15
Boyle's law, 17
buffet, 41-47
bypass air, 27
bypass duct, 27
bypass ratio, 27

C

camber, 21, 23
ceiling
    absolute, 109
    service, 109
center of gravity (CG), 177, 178
Charles's law, 17
chord, 21, 23
clearway, 66, 70
climb, 99-125
    angle of, 109-115
    at set thrust, 117
    at thrust required, 117-120
    cruise, 142-143
    economy, 120-125
    equations for, 99-100
    limitations during landing, 173
    performance with one engine inoperative,
        115-120
    rate of (*see* rate of climb)
    step, 142-143
coffin corner, 43
compressibility correction, 11-12
corrected net thrust, 29
cost index (CI), 154-155
course guidance, 93
cross-over altitude, 105
cruise, 127-152
    economy, 153-162
    effects of altitude on, 139-142
    force-speed diagrams, 127-131
    integrated range, 144-147
    integrated time, 147
    long range, 133
    maximum range, 133, 157
    specific range, 131-142
    temperature effects on, 136-137
    wind effects on, 139-142
cruise climb, 142-143
curtailed envelope, 178-185

D

density, ambient, 4
descent, rate of (*see* rate of descent)
drag, 24, 115-116
　effects of on descent, 164-165
drift-down maneuver, 165-167

E

economy cruise mach, 157-161
engine pressure ratio (EPR), 29
engines, 27
equal fuel point (EFP), 147-149
equation of state, 5, 16-17
excess power, 102

F

fan rotor rotational speed, 29-31
flaps, 22
flat-rated thrust, 33
flat rating, 33
flight management system (FMS), 153
flutter, 41
force
　friction, 96
　inertia, 16
　shearing, 16
form drag, 24
friction force, 96
fuel (*see also* cruise)
　cost index, 154-155
　endurance formulas, 151
　payload curves, 187
　point of no return and, 147-149
　range formulas, 151-152
　tankering, 149-150
fuel consumption, thrust-specific, 150
fuel consumption rate, 127
fuel flow, 31

G

generalized thrust plot, 29, 37
geopotential altitude, 6
gross gradient, 83
gross height, 84
gross path, 83
ground effect, 95
ground speed, 49

H

high-lift device (HLD), 84, 97
holding, 167-168
horizontal clearance method (HCM), 89, 91

I

improved-climb takeoff, 84, 86
induced drag, 24
inertia force, 16
International Civil Aviation Organization
　(ICAO), 1
international standard atmosphere (ISA), 1
isentropic relations, 18

K

kinetic energy rate, 102

L

landing, 172-175
　climb limitations during, 173
　maximum quick turnaround weight, 174
　maximum weight for, 177
　overweight aircraft, 175
　required distance for, 174
leading edge (LE), 21
lift, 23-24
load factor, 40-41
long range cruise (LRC), 133
long trip optimum altitude (LTOA), 144

M

Mach number, 14-15
　conversion to CAS, 10-11
　effects of, 4
　maximum operating, 46
maximum climb (MCL), 32
maximum continuous (MCT), 32
maximum cruise rating (MCR), 32, 61
maximum endurance schedule, 129-130
maximum landing weight (MLW), 177
maximum operating Mach, 46
maximum range cruise (MRC), 133, 157
maximum range schedule, 130-131
maximum takeoff weight (MTOW), 177
maximum taxi weight (MTW), 177
maximum zero-fuel weight (MZW), 177
mean aerodynamic chord (MAC), 23
mean sea level (MSL), 5
meanline, 21
motion, equations of, 168-170

N

National Oceanic and Atmospheric
　Administration (NOAA), 87
net height, 84
net path, 84
Newton's Second Law, 96

O

obstacle accountability area (OAA), 88-89
obstacles, takeoff limitations and, 83-93
optimal cruise mach, 155
outside air temperature (OAT), 2
overspeed takeoff, 84, 86

P

payload-fuel curve, 187
payload-range curve, 187-193
  equipment evaluations and, 193
pitot-static system, 7
point of no return (PNR), 147-149
position error calibration, 9
potential energy rate, 102
pressure
  ambient, 2-4
  distribution over an airfoil, 18-20
  stagnation, 4, 7
  static, 7
propulsion, 27-35

Q

quarter chord, 23

R

radius of turn, in the presence of winds, 54-58
ram air temperature (RAT), 4
rate of climb (ROC), 100-109
  acceleration factor corrections, 102-106
  effects of altitude on, 107-109
  effects of speed on, 106-107
rate of descent (ROD), 163-172
  drift-down maneuver, 165-167
  effects of drag on, 164-165
  effects of speed on, 165
  effects of thrust on, 164
  effects of weight on, 170-172
  equations of motion, 168-170
  holding and, 167-168
reference zero, 83
rejected takeoff (RTO), 67, 68-73
Reynold's number, 15-16
runways, 66-68, 80-81

S

service ceiling, 109
shearing force, 16
short trip optimum altitude (STOA), 143-144
slats, 22
sound, velocity of, 15

special takeoff procedure, 84, 87-93
specific range (SR), 131-142
speed
  compressibility correction, 11-12
  economy, 155-161
  effects of on climb, 106-107
  effects of on descent, 165
  ground, 49
  instability region, 129
  maximum, 61-62
  measurement of, 7-12
  schedules, 131
  stability during cruise, 128-129
  takeoff, 64-65
  true airspeed, 7
spoilers, 22
stagnation pressure, 4, 7
stagnation temperature, 4
stall, 41
static air temperature (SAT), 2
static pressure, 7
step climb, 142-143
stick shaker, 42
stopway, 66, 70
sweep, 23

T

takeoff (TO), 32, 63-98
  airborne portion of, 81-93
  climb limitations, 83-87
  continued, 72-76
  earthbound portion of, 67-81
  equations for, 95-98
  improved-climb, 84, 86
  limitations, 93-95
  maximum weight for, 177
  multiple V_1, 76-79
  obstacle limitations, 83-93
  overspeed, 84, 86
  rejected, 67, 68-73
  runway conditions and, 80-81
  runway configurations and, 66-68
  segments, 81-82
  special procedures, 84, 87-93
  speeds during, 64-65
  unbalanced field condition, 79-80
takeoff distance (TOD), 66
takeoff field length, 67
takeoff run (TOR), 66
takeoff weight (TOW), 63-64
tankering, 149-150
taper, 23
temperature, 15
  ambient, 2

temperature *continued*
  effects on cruise, 136-137
  outside air, 2
  ram air, 4
  stagnation, 4
  static air, 2
  total air, 4
terminal instrument procedures (TERPS), 88
thrust
  available, 37
  effects of on descent, 164
  flat-rated, 33
  limitations, 34-35
  required, 37-40
thrust ratings, 31-34
thrust-setting parameters (TSP), 29
thrust-specific fuel consumption (TSFC), 150
top of climb (TOC), 163
top of descent (TOD), 163
total air temperature (TAT), 4
trailing edge (TE), 21
true airspeed (TAS), 7
turbofans, 27
turbojets, 27

turning (*see also* bank-angle)
  in the presence of winds, 54-58

V
V-*n* diagrams, 58-61
velocity, 151
velocity of sound, 15
vertical clearance method (VCM), 89-91
viscous drag, 24

W
wave drag, 24
weight
  balance and, 177-186
  landing, 177
  maximum takeoff, 177
  maximum taxi, 177
  operational allowances and curtailed
    envelope, 178-185
  payload-range curves, 187-193
  zero-fuel, 177
wind, effects on cruise, 139-142
wind shift function (WSF), 140
wings, properties, 22-23

About the author

Carlos Padilla has aerospace engineering degrees from Texas A&M University and has practiced in the field for 20 years in the United States and abroad. During the past 10 years, he has been involved in many aspects of aircraft performance engineering both as a practitioner and as a software designer.

CPSIA information can be obtained at www.ICGtesting.com
Printed in the USA
BVOW09s1922040216

435557BV00009B/59/P